A *Promenade* Along Electrodynamics

Junichiro Fukai
Department of Physics
Auburn University

VALES LAKE PUBLISHING, LLC

Library of Congress Control Number: 2002117784

Fukai, Junichiro
A Promenade Along Electrodynamics
Howard C. Hayden, Editor
Includes bibliographic references and index
1. Physics
2. Electricity
3. Magnetism
4. Electrodynamics
5. Relativity
6. Ampere's Law
7. Weber's Law
8. Maxwell's equations

ISBN 0-9714845-1-1

Copyright © 2003 by Junichiro Fukai
All Rights Reserved.
Printed in the USA

This book or any part thereof must not be reproduced by microfilm, photostat, scanner, or any other process without written permission of the publisher.

Cover design by Chris Mascarenas

$21.95 from
Vales Lake Publishing, LLC
P.O. Box 7595
Pueblo West, CO 81007-0595
SAN: 2 5 4 - 2 5 3 6

Acknowledgements

I wish to thank my colleagues at Auburn University, particularly Dr. Earl T. Kinzer. Special mention must be made of Professor Andre K. T. Assis of the State University of Campinas, Brazil, Dr. Edward G. Harris, Professor Emeritus of the University of Tennessee, and Dr. Howard C. Hayden, Professor Emeritus of the University of Connecticut for countless helpful remarks and discussions. My thanks are also extended to Professor Joseph D. Perez of Auburn University for reading and commenting on a preliminary version of the manuscript.

Junichiro Fukai
Auburn, Alabama, January 2003

Preface

For what purpose has this book been written? One thing for sure is that this book is not intended to be a textbook. My intention is to sketch, in a broad outline, the attempts of the human mind to find a connection between the world of ideas and the world of phenomena. I am trying to show how different ideas explain the world of electromagnetic phenomena. Suppose that you want to go to a destination. A drive on a highway with clear signs has a predictable view. If you take a detour, and, say, go around a mountain or go through a forest, you might encounter unexpected beauty. This book is about a *promenade* a little off the beaten path in electrodynamics. It is a simple chat among all of us, as students of science. My aim is to clarify the present paradigm of electromagnetic theory and give the reader some insight into the eternal struggle within the inventive human mind for a fuller understanding of the laws governing electromagnetic phenomena.

While I was studying the forces on a charged particle in a magnetic field produced by another charged particle in motion, I had difficulty picturing their mutual interactions. There are only two particles in relative motion. According to the action-reaction principle of Newton's third law, the forces on each particle must be equal and opposite. Formulas in an elementary textbook, however, give a conclusion that a view by one particle is different from that of the other particle. I understood later that here we need Einstein's special theory of relativity to fix this paradox. But I intended to explore the possibility for an alternate explanation. This book is about the story of how, during my intuitive endeavor of investigation, I encountered forgotten theories such as Ampere's force law derived in 1823 [Andre-Marie Ampere, 1775-1836], and Weber's electrodynamics [1], developed around 1848 [Wilhelm Eduard Weber, 1804-1891]. Weber's theory is an interesting one. It is based on the principle of action at distance, thus, directly satisfying Newton's third law.

In Weber's electrodynamics only the charges, their separations, their velocities (the first time derivatives) and accelerations (second time derivatives) are important to describe the forces between moving charges. Historically, studies based on the action at distance were in the main train of thought on the Continent. British physicist James Clerk Maxwell [1831-

1879] had different conceptions. He liked the concept of fields, a visualization of the interaction of two charges, proposed by Michael Faraday [1791-1867]. Based on the ether and fields, he showed that his model could also explain the main phenomena associated with electromagnetism. He argued that all electromagnetic phenomena could be explained as sorts of strains and rotational motions of ether. The permittivity and permeability were introduced for ether. Furthermore, Maxwell was able to predict the existence of electromagnetic wave. He considered electromagnetic waves as traveling disturbances of fields in ether. When Heinrich Rudolf Hertz [1857-1894] experimentally verified the existence of electromagnetic waves, people thought that the existence of ether was also proved. However, in 1887 the Michelson-Morley [Albert Abraham Michelson, 1852-1931; Edward Williams Morley, 1838-1923] experiments raised serious doubt against the existence of ether. In addition, the theories based on electric and magnetic fields had some deficiencies in describing effects between moving charges. Albert Einstein [1879-1955] tried to fix the shortcomings by his special theory of relativity based on the postulate that no ether exists and the speed of light is a universal constant. As the special theory of relativity became widely accepted, so did Maxwell's electromagnetic theory without ether. Together the two theories became the paradigm for the study of electromagnetic phenomena. The concept of the field established for itself a leading place in physics and has remained one of the basic physical concepts. The electromagnetic fields are, for modern physicists, as real as the chair on which they sit. Now we only talk about fields, associated potentials, and delayed action due to finite velocity of propagation of the interactions.

On the other hand, Weber's theory nowadays is sometimes considered "erroneous" [2], or it is regarded heretical [3], even though Weber's theory was quite successful in correctly describing the known phenomena of electromagnetism.

In science, heresy or heretical theory is considered to contain some kind of error, mistake, delusions, fallacy, or unscientific argument. Those errors have several patterns: an error that arises owing to the generalization of phenomenon through direct observational experiences, e.g., Ptolemaic system (earth-centered solar system); an incorrect representation of an object introducing a nonexistent entity, e.g., ether; a mistake caused by subjective observations due to human psychological conviction, e.g., N-rays; and mistakes caused by extending an application beyond the limitation fundamental theory. There are also many common mistakes due to unscientific procedures, or incorrect analysis.

Once a theory is regarded heretical, a paper dealing with discussions based on the theory will not be accepted nor even reviewed by most of major refereed journals. Weber's theory became heretical, even though its conceptual framework permits the explanation and investigation of the known electromagnetic phenomena. It is interesting to note that Weber's theory is recently shown to be compatible with Maxwell's theory since both contain the basic laws of Ampere and Faraday [4]. Weber's theory cannot be called wrong in the scientific term. The conceptual framework of the theory does not have any logical error. Weber's theory could show that electrical signals propagate at the speed of light in a conducting wire. Weber's theory, however, is apparently incomplete, in that it supposedly cannot predict the propagation of electromagnetic waves in free space. This is considered to be the crucial defect of Weber's theory. On this issue, I will show that Weber's theory can predict the propagation of electrical signals in free space at the speed of light if we borrow a modern quantum mechanical description of the vacuum (Chapter 9).

The book starts with an introduction describing my motivation for this endeavor and a review of elementary electricity and magnetism in Chapter 1. In Chapter 2, we will derive a new formula for the forces between two charges moving at constant velocities. In Chapter 3, by using the new force formula, we will consider several electromagnetic situations of two charges in relative motion, then compare the results with those of the conventional theory including Einstein's special theory of relativity. Chapter 4 will show that the new formula leads to a force formula for the interaction of two current elements, so called Ampere's force law. In Chapter 5 we will extend calculation for a formula for the forces of two charges moving at a high speed. We will discuss the similarities and discrepancies between our results and the conventional ones. In Chapter 6, by using the high-speed force formula, we can show a relativistic effect for Rutherford scattering. Chapter 7 deals with a more rigorous treatment by including their relative acceleration between two charges in relative motion. Here, we will rediscover

Weber's force law for two moving charges. In Chapter 8 we show that Weber's law can be easily applied to problems that are seemingly complicated in an orthodox view. In this Chapter we will also derive some effects that are not predicted by Maxwell's theory but are predicted by Weber's electrodynamics. We will propose some experiments to test these predictions. In Chapter 9 we will discuss possibilities of electrical signals traveling in free space at the speed of light. In order to do so, we require the modern view of the vacuum. Also from the modern view of the vacuum, we will present a simple explanation of bending light around a massive star without using the general theory of relativity. Finally, we will discuss a further development of Weber's theory into other aspects of physics.

In this book I am not attempting to say that Weber's theory is better than Maxwell's theory. Given the extreme scarcity of easily available publications about Weber's theory, my intention is reacquaint the scientific community with the concept of unjustly undervalued theory in the hope that it will provide a stimulus for a better understanding of difficult electromagnetic phenomena.

A table of notable physicists in the historical order who appear in this book is provided in Appendix A. Their brief biographies are presented in Appendix B.

TABLE OF CONTENTS

ACKNOWLEDGEMENTS III

PREFACE V

TABLE OF CONTENTS IX

CHAPTER 1. INTRODUCTION 1
    MOTIVATION 2
    REVIEW OF ELECTRICITY AND MAGNETISM 4

CHAPTER 2. FORCES BETWEEN TWO CHARGES MOVING WITH CONSTANT VELOCITIES 11
    DERIVATION OF A "NEW" FORMULA TO REPLACE LORENTZ'S FORCE 11

CHAPTER 3. COMPARISON WITH RELATIVISTIC TREATMENT 27
    CASE 1 - TWO CHARGES MOVING WITH THE SAME VELOCITY 27
    CASE 2 - ONE CHARGE STATIONARY AND THE OTHER MOVING 30
    CASE 3 - TWO CHARGES MOVE PERPENDICULAR TO EACH OTHER 31
    CASE 4 - TWO CHARGES MOVE ON A STRAIGHT LINE 34

CHAPTER 4. AMPERE'S FORCE LAW 37
    DERIVATION OF AMPERE'S FORCE LAW 37

CHAPTER 5. HIGH SPEED LIMIT AND COMPARISON WITH THE SPECIAL THEORY OF RELATIVITY 43
    DERIVATION OF HIGHER-ORDER TERM CORRECTIONS 43

CHAPTER 6. SCATTERING AND ORBIT PROBLEM 49
    SCATTERING 49
    ORBIT PROBLEM 52

CHAPTER 7. A MORE GENERAL THEORY — AN ENCOUNTER WITH WEBER'S FORCE LAW 55
WEBER'S LAW 59
COMPATIBILITY WITH MAXWELL'S THEORY 62

CHAPTER 8. APPLICATIONS OF WEBER'S ELECTRODYNAMICS 67
TROUTON-NOBLE EXPERIMENT 67
ROTATING SOLENOID 68
INTERACTION OF TWO MOVING CHARGES 72
INERTIAL MASS OF A CHARGED PARTICLE IN A POTENTIAL 73
WEBERIAN INDUCTION 75

CHAPTER 9. THE PROPAGATION OF ELECTRICAL SIGNALS IN VACUUM 79
REVIEW OF VACUUM 79
ELECTRIC SIGNAL PROPAGATION IN VACUUM 81
BENDING OF LIGHT NEAR A MASSIVE BODY 82

EPILOGUE 89
CONCLUDING REMARKS 89
FURTHER IMPLICATIONS 91

APPENDIX A: NOTABLE PHYSICISTS ON ELECTRICITY AND MAGNETISM IN THE 19TH CENTURY 95

APPENDIX B: BIOGRAPHIES OF NOTABLE PHYSICISTS IN THE 19TH CENTURY 97

REFERENCES 119

INDEX 125

ABOUT THE AUTHOR 127

Chapter 1. Introduction

When we try to solve a problem, we sometimes encounter a situation that gives one answer when analyzed one way, and a different answer when analyzed another way — a paradox or confusion. Confusion that we talk about is concerned with the expression of the force on a charge in a magnetic field proposed by Dutch physicist Hendrik Autoon Lorentz [1853-1928]

$$F = q(E + v \times B)$$ Lorentz's Force

where q is the charge experiencing the force, v is its velocity and E and B are electric and magnetic fields. The electric field E and magnetic field B are examples of force fields that arise when interacting forces are acting over a distance.

Many common forces might be referred to as "contact forces," such as the force you push a cart and the force a baseball bat exerts on a ball. On the other hand, both gravitational force and the electrical force act over a distance: there is a force even when the two objects are not in contact. Why is the body able to exert a force on a body that is separated in space? This difficulty can be overcome with the idea of the field, developed by the British scientist Michael Faraday [1791-1867].

In the electrical case, according to Faraday, an electric field extends outward from every charge and permeates all of space. When a second charge is placed near the first charge, it feels a force because of the electric field that is there. The empty space is somehow altered to exert a force on a charge there. The electric field at the location of the second charge is considered to interact directly with this charge to produce the force. You might say that the field acts as "go-between." A charge generates an electric field that permeates the surrounding space and exerts a force on another charge that it touches. Through the field, the electric interaction between charges is now understandable as action-by-contact. A field may be considered as an alteration of space such that a body there experiences a force. If the force is gravitational, then there exists a gravitational field and if the force is magnetic force then there exists a magnetic field. It must be emphasized, however, that a field is not a kind of matter. It is, rather a concept, though some think the field is real. The magnetic field can be considered as an alteration of space due to a moving charge. A moving charge or an electric current creates a magnetic field according to Eq. (1-8) or (1-13) that will be discussed later in this chapter.

Suppose that somehow a magnetic field is produced in space. What does this field do on an electric charge coming in the region? If a charge q moving with velocity **v** in a region of magnetic field **B** will experience a magnetic force given by $q\mathbf{v} \times \mathbf{B}$, that is, the magnitude of the force is $qvB\sin\theta$ where θ is the angle between the direction of the velocity and that of the magnetic field, and the direction of the force is perpendicular to both **v** and **B**. What is this velocity **v**? Obviously, velocity is not an intrinsic property of a body. Velocity is a relation between the charge and a certain body relative to which it is moving.

Unfortunately, most textbooks do not specify the velocity relative to what. Facing to this indefiniteness the student often becomes confused between several possibilities: velocity of charge relative to the magnet or a current carrying wire that generates **B**; relative to the laboratory; relative to any observer or frame of reference; relative to the magnetic field; relative to the drift velocity of the current producing charges that generate **B**; relative to the **B**-field detector and so on. Historically, there was no need to say what velocity meant. A charge moves through the laboratory where the observer sits and this apparatus also sits. The velocity was the velocity of the charge with respect to the laboratory. In considering the theoretical force between relatively moving charges the question did not come up. We have to wait until the introduction of the special theory of relatively to clearly state the velocity as the velocity of a charge relative to an observer, or his inertial frame of reference. We would wonder why a force depends upon the observer at all. The interaction is between a charge q and the charges that are the sources of **E** and **B**, not charges and the observer. Keep this in mind and later in Chapter 3 we will treat the relativistic aspect of the interaction of two charges in motion.

Motivation

My detour to this endeavor of electrodynamics started when I encountered a paradox or confusion that we will describe in the following (Fig. 1-1) [5].

We look at the force on q_1 due to q_2 and vice versa. On q_2 only the electric force from q_1, since q_1's magnetic field has no component along its line of motion. On q_1, however, there is again the electric force, in addition, a magnetic force, since it is moving in a magnetic field made by q_2. The forces are drawn in Fig 1-1 (b). The electric forces on q_1 and q_2 are equal and opposite. However, there is sideways (magnetic) force on q_1 and no sideways force on q_2. Action does not equal reaction! The result depends on the coordinate system we choose. The dependence on the coordinate system contradicts the assumption of classical mechanics that forces are the same in all inertial coordinate systems. The reader can check the forces on the above charges by using the formulas given by Eq. (1-11) with Eq. (1-12) and Eq. (1-13) that are developed in the following.

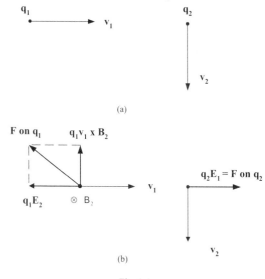

Fig. 1-1.
(a) Two charges move with velocities at right angles.
(b) The forces between two moving charges are not always equal and opposite.

This apparent paradox gives a hint that some remedy is necessary to resolve the inconsistencies. The remedy proposed by Einstein is the special theory of relativity. A proper application of the relativistic transformations between coordinates of **E** and **B** can resolve the paradox. We will discuss this subject at a later chapter.

Our natural question may be "Can we develop a theory that formulates the forces on two charges are equal and opposite and that satisfies Newton's third law?" We are going to seek such a theory without relying on the special theory of relativity. In Chapter 2, we will develop a "new formula" that just describes the required forces.

Before we go further into our adventure, let us first review electricity and magnetism in the very beginning level.

Review of Electricity and Magnetism

We are familiar with magnets and have marveled at a magnet to attract nails and other iron and steel objects. The class of phenomena involving magnetic forces is called magnetism. Magnetism is known and studied from the early days of civilization, as is electricity. In the 16th century, William Gilbert made artificial magnets by rubbing pieces of iron against lodestone, and suggested that a compass always points north and south because the earth itself has magnetic properties. In 1750, John Michell found the magnetic poles obey the inverse square law just like the force between two electric charges, and his results were confirmed by Charles Coulomb [1736-1806]. Until the 19th century, the subjects of magnetism and electricity developed independently of each other. One important difference between them is that there exist no separate magnetic charges (single monopoles) in magnetism. We now know that magnetism is closely related to electricity.

In the early 19th century, Danish physicist Hans Christian Oersted [1777-1851] discovered that an electric current flowing through a wire deflects a magnetic needle placed in its neighborhood in such a way that the needle assumes a position perpendicular to the plane passing through the wire and through the center of the needle. When we replace the needle with another current carrying wire placed parallel to the original wire, the wires attract each other if the currents are in the same direction and repel each other if the currents are in opposite directions. The force between two parallel wires carrying currents I_1 and I_2 is proportional to $I_1 I_2 / r$, where r is the distance between the wires. The forces coming here are magnetic forces. All magnetic effects can ultimately be traced to electric currents, or to moving charges.

The following discussions are as standard as those that will be found on a standard textbook [6-8].

In the general case of a pair of currents as shown in Fig. 1-2, the magnetic force which one current exerts on the other when both are in free space is given by

INTRODUCTION

$$\mathbf{F}_{12} = \frac{\mu_o}{4\pi} I_1 I_2 \oint_1 \oint_2 \frac{d\mathbf{l}_1 \times (d\mathbf{l}_2 \times \hat{\mathbf{r}})}{r^2} \qquad (1\text{-}1)$$

where $\mathbf{F}_{12}$ is the force exerted on current I_1 by current I_2, and where the line integrals are evaluated over the two wires. This is the magnetic force law.

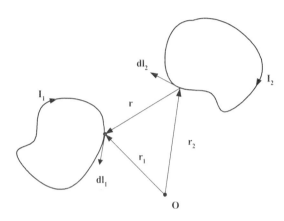

Fig. 1-2

The vectors $d\mathbf{l}_2$ and $d\mathbf{l}_1$ point in the directions of positive current flow, $\mathbf{r}$ ($= \mathbf{r}_1 - \mathbf{r}_2$), which is the displacement vector pointing from $d\mathbf{l}_2$ located at $\mathbf{r}_2$ to $d\mathbf{l}_1$ at $\mathbf{r}_1$ and r is the distance between the two current elements $d\mathbf{l}_2$ and $d\mathbf{l}_1$. The constant μ_o is called the permeability of free space (vacuum) and its value is $4\pi \times 10^{-7}$ N/A^2 in the SI (MKS) system.

The force $\mathbf{F}_{12}$ is expressed in Eq. (1-1) does not appear symmetrical between two currents. This is quite disturbing, since, from Newton's third law, we expect $\mathbf{F}_{12}$ to equal $-\mathbf{F}_{21}$. The force can be expressed in a symmetrical way by manipulating the integral of Eq. (1-1). We can show the result as

$$\mathbf{F}_{12} = -\frac{\mu_o}{4\pi} I_1 I_2 \oint_1 \oint_2 \frac{\hat{\mathbf{r}}(d\mathbf{l}_1 \cdot d\mathbf{l}_2)}{r^2} \qquad (1\text{-}2)$$

We now have $\mathbf{F}_{12} = -\mathbf{F}_{21}$, since the unit vector $\hat{\mathbf{r}} = \mathbf{r}/r$ is directed toward the circuit on which the force is to be calculated. Newton's third law therefore applies.

The expression of Eq. (1-1), however, has a desired form to separate the part of circuit 1 from that of circuit 2 in the double integration. We can express it as the interaction of current 1 with the field of current 2. We can perform such an operation on Eq. (1-1) and obtain

$$\mathbf{F}_{12} = I_1 \oint_1 d\mathbf{l}_1 \times \left(\frac{\mu_0}{4\pi} I_2 \oint_2 \frac{d\mathbf{l}_2 \times \hat{\mathbf{r}}}{r^2} \right) \tag{1-3}$$

or,

$$\mathbf{F}_{12} = I_1 \oint_1 d\mathbf{l}_1 \times \mathbf{B} \tag{1-4}$$

where

$$\mathbf{B} = \frac{\mu_o}{4\pi} I_2 \oint_2 \frac{d\mathbf{l}_2 \times \hat{\mathbf{r}}}{r^2} \tag{1-5}$$

can be taken as the field of circuit 2 (or current 2) at the position of the element $d\mathbf{l}_1$ of circuit 1. Note here that obtaining Eq. (1-3) is to separate the contribution of circuit 1 and circuit 2 in the integrand of Eq. (1-2) so that circuit 2 explicitly gives the magnetic force on the circuit 1. The expression **B** given by Eq. (1-5) is called the magnetic field due to current 2 at a displacement **r**. In general, we may omit the subscript 2 and write, for the magnetic field produced by a current circuit,

$$\mathbf{B} = \frac{\mu_o}{4\pi} I \oint \frac{d\mathbf{l} \times \hat{\mathbf{r}}}{r^2} \qquad \text{Biot-Savart law} \tag{1-6}$$

This expression for **B** is often called the Biot-Savart law.

The element of force $d\mathbf{F}$ on an element of wire of length $d\mathbf{l}$ carrying a current I in a region where the magnetic field is **B** is then given by, from Eq. (1-4),

$$d\mathbf{F} = I d\mathbf{l} \times \mathbf{B} \tag{1-7}$$

In a similar fashion, the element of magnetic field $d\mathbf{B}$ produced by an element of wire of length $d\mathbf{l}$ carrying a current I can be written as

INTRODUCTION

$$d\mathbf{B} = \frac{\mu_o}{4\pi} \frac{I d\mathbf{l} \times \hat{\mathbf{r}}}{r^2} \qquad (1\text{-}8)$$

We wish now to rewrite the above formulas for the interaction of two charges in motion, instead of the current elements.

An electrical current in a conducting wire is electrically neutral since there are equal numbers of positive and negative charges. We can neglect the contribution of the electric Coulomb force when we consider the interaction of two neutral current elements. In case of two moving charges, however, the electrostatic Coulomb force comes to play. The electrical force between them is much stronger than the magnetic force. French physicist Charles Coulomb [1736-1806] established the electrical force law between two stationary charges, in 1787, and the force on q_1 is given by

$$\mathbf{F}_e = \frac{1}{4\pi\varepsilon_o} \frac{q_1 q_2}{r^2} \hat{\mathbf{r}} \qquad \text{Coulomb's law} \qquad (1\text{-}9)$$

where $\hat{\mathbf{r}}$ is the unit vector of $\mathbf{r} = \mathbf{r}_1 - \mathbf{r}_2$ and ε_o is called the permittivity of free space (vacuum) and its value is 8.85×10^{-12} farad/m in SI units.

We need now a formula for the magnetic force between moving charges that is applicable for cases such as given in Fig. 1-1. We can do so by modifying Eq. (1-1), valid for current segments, into an expression for the currents of moving charges. Assume that two charges q_1 and q_2 are moving at $\mathbf{v}_1$ and $\mathbf{v}_2$ respectively (Fig. 1-3).

The current elements $I_1 d\mathbf{l}_1$ and $I_2 d\mathbf{l}_2$ of Eq. (1-1) may be replaced by $q_1 \mathbf{v}_1$ and $q_2 \mathbf{v}_2$ respectively [6-8]. Thus, we can rewrite Eq. (1-1) for the magnetic force on q_1 exerted by q_2 as

$$\mathbf{F}_m = \frac{\mu_o}{4\pi} \frac{q_1 q_2}{r^2} \left[\mathbf{v}_1 \times (\mathbf{v}_2 \times \hat{\mathbf{r}}) \right] \qquad (1\text{-}10)$$

where we again used $\hat{\mathbf{r}}$, the unit vector of $\mathbf{r}$. These forces are combined

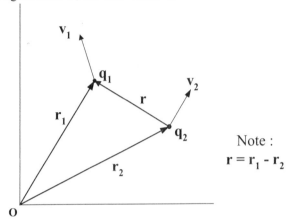

Fig. 1-3

and the total force on q_1 due to q_2 is written by

$$\mathbf{F} = q_1(\mathbf{E} + \mathbf{v}_1 \times \mathbf{B}) \quad \text{Lorentz's force} \quad (1\text{-}11)$$

where the electric field $\mathbf{E}$ due to q_2 is defined as

$$\mathbf{E} = \frac{1}{4\pi\varepsilon_o} \frac{q_2}{r^2} \hat{\mathbf{r}} \quad (1\text{-}12)$$

and the magnetic field $\mathbf{B}$ due to the motion of q_2 is defined as

$$\mathbf{B} = \frac{\mu_o}{4\pi} \frac{q_2}{r^2} (\mathbf{v}_2 \times \hat{\mathbf{r}}) \quad (1\text{-}13)$$

We note that the constants μ_o and ε_o are related, and the product of two has the value of $1/c^2$ where c is the speed of light in vacuum. Using $\varepsilon_o\mu_o = 1/c^2$, Eq. (1-13) can be written

$$\mathbf{B} = (1/c^2)\, \mathbf{v}_2 \times \mathbf{E}_2 \quad (1\text{-}14)$$

where $\mathbf{E}_2$ is the electric field due to a source charge q_2 with velocity $\mathbf{v}_2$.

It is remarkable that the magnetic force does not depend only on the relative velocity of the two charges, but is different in a moving coordinate system. The expression of Eq. (1-10) is not symmetrical, that is, the magnetic force simply does not change sign when the particle labels are exchanged. In particular it vanishes in a coordinate system moving with the source particle. The formalism above contradicts with the assumption of classical mechanics that forces are the same in all inertial coordinate systems. Here we see that a new theory is needed to fix the contradiction. Albert Einstein in 1905 developed the special theory of relativity to accommodate Lorentz's force to be consistent with mechanics. We will put aside the special theory of relativity for a while. For the above observation, some authors emphasize that Eq. (1-13) is only true for particles moving with velocities small compared to that of light [9], or that the expression of Eq. (1-13) is simply wrong and is only approximately right for negligible relativistic effects [10], or omit entirely mentioning an expression for a magnetic field created by a moving charge. The magnetic effects by a moving charge can arise for any value of their velocity [11]. We will show later that if we replace the velocity v_2 of Eq. (1-14) by the relative velocity between moving charges the expression should be valid for classical cases. With this revised view of Eq. (1-14), the basic idea of the fine structure of the hydrogen atom can be, for example, effectively explained.

How can we fix this asymmetric nature of the magnetic force expressed in Eq. (1-10)? We wonder what kind of formalism can lead to the formula in which the resulting forces are equal and opposite from each other. What we are interested in here is finding such a formula, without relying on the special theory of relativity. One crucial feature we require is that we must work in the frame of two charges that are in relative motion.

Chapter 2. Forces between two charges moving with constant velocities

Since our confusion started with the application of Lorentz's force, our challenge is to find a "new" formula to replace Lorentz's formula so that we can correctly predict how the two moving charges exert forces on each other. Our discussions will be limited to theoretical endeavor.

Our development here, though, should not go beyond the scope of the well-established Maxwell's equations. We assume that Maxwell's equations are valid and consistent with experimental findings in the classical regime. We are only looking for a formula to replace Lorentz's expression with one that does not contradict Newton's third law. In the following, we will develop a treatment for the interaction of two moving charges from only a relative perspective from each other. It will turn out to be that the resulting formula will lead to an old forgotten force formula derived by the French physicist Andre-Marie Ampere [1775-1836].

Derivation of a "new" formula to replace Lorentz's force

Consider a source charge q_s, located at $\mathbf{r}_s$ and a receiver test charge q_r located at $\mathbf{r}_r$ in a given coordinate system. The source charge and the receiver charge have velocities, $\mathbf{v}_s$ and $\mathbf{v}_r$, respectively (Fig. 2-1).

We assume here that the velocities are constant. A more general treat-

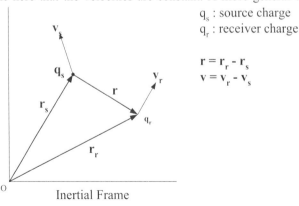

Fig. 2-1

ment with accelerations will be discussed in Chapter 7. The relative position vector, $\mathbf{r} = \mathbf{r}_r - \mathbf{r}_s$, and relative velocity vector $\mathbf{v} = \mathbf{v}_r - \mathbf{v}_s$ are those of q_r viewed from q_s at the origin.

The problem we will discuss here is how the receiver charge will experience a force due to the source charge in relative motion. The relative velocity of the source charge from the viewpoint of the receiver charge is $\mathbf{u} = \mathbf{v}_s - \mathbf{v}_r$ ($= -\mathbf{v}$). We will assume that when at rest, the force on either due to the other will be given by Coulomb's force law. Faraday's law and Ampere-Maxwell equation will be assumed to be valid in their source-free forms as experiments validate them.

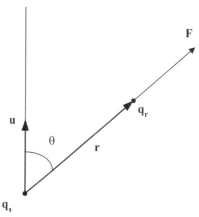

Fig. 2-2

We will set up a polar coordinate system in which the relative velocity vector, $\mathbf{u}$, determines the polar axis and angle θ. In Fig. 2-2, q_s is at $\mathbf{r} = 0$ and q_r is the subject (receiver) charge at $\mathbf{r}$. The velocity of the source charge relative to q_r is $\mathbf{u}$ and θ is the angle between $\mathbf{u}$ and $\mathbf{r}$. The static force on q_r due to q_s is $\mathbf{F} = q_r q_s \mathbf{r}/(4\pi\varepsilon_0 r^3)$.

The essential idea that we will use to guide us, is that $\mathbf{F}$ can be a manifestation of a static electric field $\mathbf{E} = \mathbf{F}/q_r$ and, if q_s and its field $\mathbf{E}$ are moving (quasistatically) with respect to q_r, then a magnetic field exists in the neighborhood of q_r and will change in time due to the motion of q_s and its field $\mathbf{E}$. This changing magnetic field will, in turn, result in an induced electric field that may be interpreted as a velocity-dependent correction to $\mathbf{E}$ at the position of q_r. The resulting position- and velocity-dependent force on

FORCES BETWEEN MOVING CHARGES

q_r will thus be entirely electric in nature, rather than a combination of electric (static) and magnetic (dynamic) forces.

In proceeding, we will not assume the validity of either the Biot-Savart or Lorentz force relations, but rely entirely on the Ampere-Maxwell and Faraday equations. The Ampere-Maxwell equation will be used in integral form as

$$\oint_{c(s)} \mathbf{B} \cdot d\mathbf{l} = \frac{1}{c^2} \frac{d}{dt} \iint_S \mathbf{E} \cdot \hat{\mathbf{n}} dA \qquad (2\text{-}1)$$

where $c(s)$ is the boundary of the surface S and $\mathbf{n}$ is the surface unit vector. The **B** field is assumed to have axial symmetry with respect to **u**, and $\mathbf{B} = B\hat{\phi}$, i.e., the lines of **B** will be circles centered on the polar axis. Fig. 2-3 shows that the surface S is a spherical segment centered on q_s and with the boundary passing through the location of q_r.

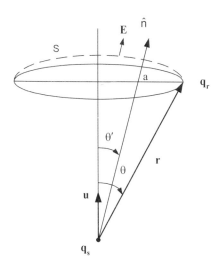

Fig. 2-3

The electric flux through S of the right-hand side of Eq. (2-1) is

$$\iint_s \frac{q_s dA}{4\pi\varepsilon_o r'^2} = \frac{q_s}{4\pi\varepsilon_o} \int_0^\theta \int_0^{2\pi} \frac{r'^2 \sin\theta' d\theta' d\phi'}{r'^2} \tag{2-2}$$

and the line integral around $c(s)$ of the left-hand side of Eq. (2-1) is

$$\oint_{c(s)} \mathbf{B}\cdot d\mathbf{l} = B 2\pi a = 2\pi r B \sin\theta \tag{2-3}$$

Thus, we find

$$B = \frac{q_s}{4\pi\varepsilon_o c^2 r} \frac{d\theta}{dt} \tag{2-4}$$

Now, by considering the geometry from the point of view of q_s, we have $\mathbf{u} = (u\cos\theta)\hat{\mathbf{r}} - (u\sin\theta)\hat{\boldsymbol{\theta}}$. Since $-\mathbf{u}\cdot\hat{\boldsymbol{\theta}} = u\sin\theta = r(d\theta/dt)$ and $-\mathbf{u}\cdot\hat{\mathbf{r}} = dr/dt = -u\cos\theta$, Eq. (2-4) becomes

$$B = \frac{q_s u \sin\theta}{4\pi\varepsilon_o c^2 r^2} \tag{2-5}$$

Since **B** is in the ϕ direction and the static electric field $\mathbf{E} = q_s \mathbf{r}/(4\pi\varepsilon_0 r^3)$, this can be written as

$$\mathbf{B} = \frac{1}{c^2} \mathbf{u} \times \mathbf{E} \tag{2-6}$$

This expression may appear to be identical to the Biot-Savart relation of Eq. (1-14). They are different. The difference is that Eq. (2-6) is due to q_s with a *relative velocity*, **u** and it does not depend on the choice of a coordinate system. If $\mathbf{v}_s = 0$ in a coordinate system, the magnetic field vanishes according to the Biot-Savart law of Eq. (1-14), but the new expression of Eq. (2-6) gives non zero values as long as **u** is not zero, just as experiments indicate.

The magnetic field of Eq. (2-6) or Eq. (2-5) changes with time since the radial distance r and the angle θ change as the source charge moves. We will seek the induced electric field arising from this change of the magnetic field. It is convenient to use Faraday's law in differential form

FORCES BETWEEN MOVING CHARGES

$$\nabla \times \mathbf{E} = -(d\mathbf{B}/dt)$$

From Eq. (2-5), and noting that $\mathbf{B} = B\hat{\varphi}$ and $B = B(r, \theta)$, we have

$$\frac{1}{r}\frac{\partial}{\partial r}(rE_\theta) - \frac{1}{r}\frac{\partial E_r}{\partial \theta} = -\frac{3q_s u^2 \sin\theta \cos\theta}{4\pi\varepsilon_o c^2 r^3} \tag{2-7}$$

where E_r and E_θ are the radial and the θ direction components of induced electric field respectively. For non-zero values of r, we obtain

$$\frac{\partial}{\partial r}(rE_\theta) - \frac{\partial E_r}{\partial \theta} = -\frac{3q_s u^2 \sin\theta \cos\theta}{4\pi\varepsilon_o c^2 r^2} \tag{2-8}$$

Now we must solve the above differential equation for the two unknown quantities, E_r and E_θ. An additional relation involving can be found from $\nabla \cdot \mathbf{E} = 0$ except at $r = 0$. The divergence of $\mathbf{E}$ is zero since q_s is the only source for the field on q_r. This gives

$$\frac{1}{r^2}\frac{\partial}{\partial r}(r^2 E_r) + \frac{1}{r\sin\theta}\frac{\partial}{\partial \theta}(E_\theta \sin\theta) = 0 \tag{2-9}$$

So, we have a pair of coupled partial differential equations, Eq. (2-8) and Eq. (2-9) to solve for E_r and E_θ. We will consider the second of these relations first. It is intuitively clear that in Euclidian space $\mathbf{E}$ (and $\mathbf{F}$) should, for a point or localized source, diverge no more strongly that $1/r^2$. Also, there is no indication in either of the two equations that this notion is in error. Accordingly, we may assume that $\partial(r^2 E_r)/\partial r = 0$. The remaining term then indicates that $(E_\theta \sin\theta)$ is independent of θ. This requires that E_θ must vanish. Furthermore, if $\mathbf{E}$ (and $\mathbf{F}$) had a θ component, then we would have the unphysical consequence of a net internal torque for the system of an isolated pair of two particles. Therefore we expect $E_\theta = 0$ and seek a radial $\mathbf{E}$.

We then must solve for E_r by

$$\frac{\partial E_r}{\partial \theta} = \frac{3q_s u^2 \sin\theta \cos\theta}{4\pi\varepsilon_o c^2 r^2} \tag{2-10}$$

Solving for E_r,

$$E_r = \int \frac{3q_s u^2 \sin\theta \cos\theta}{4\pi\varepsilon_o c^2 r^2} d\theta \, d\theta \tag{2-11}$$

Introducing $f(r)$, an integration constant that is independent of θ, we have

$$E_r = \frac{3q_s u^2 \sin^2\theta}{8\pi\varepsilon_o c^2 r^2} + f(r) \tag{2-12}$$

Now E_r should reduce to the Coulomb law when $u \to 0$, and for $f(r)$, we include the possibility of other θ independent terms. We write

$$E_r = \frac{q_s}{4\pi\varepsilon_o r^2}\left(1 + \frac{3}{2}\frac{u^2}{c^2}\sin^2\theta + K\frac{u^2}{c^2}\right) \tag{2-13}$$

with K to be determined.

Then, the force on q_r due to q_s is given by

$$\mathbf{F} = \frac{q_s q_r}{4\pi\varepsilon_o r^2}\left(1 + \frac{3}{2}\frac{u^2}{c^2}\sin^2\theta + K\frac{u^2}{c^2}\right)\hat{\mathbf{r}} \tag{2-14}$$

The expression is obviously symmetric between q_s and q_r and satisfy Newton's third law. The additive arbitrary function of r and u is to be determined subsequently.

The preceding expression of the force between q_s and q_r is derived by computing the magnetic flux change at the position of the receiver charge as the source charge moves at a velocity $\mathbf{u}$. It is instructive to manipulate Faraday's law from the viewpoint of the receiver charge that is moving with velocity, $-\mathbf{u}$, through the $\mathbf{B}$ pattern associated with q_s. We should obtain the same result. Since the change of $\mathbf{B}$ viewed by q_r is the total time derivative of $\mathbf{B}$, i.e., Faraday's law is

$$\nabla\times\mathbf{E} = -\frac{d\mathbf{B}}{dt} = -\left(\int\frac{\partial\mathbf{B}}{\partial t} + \mathbf{v}\cdot\nabla\mathbf{B}\right) = -\frac{\partial\mathbf{B}}{\partial t} + \mathbf{u}\cdot\nabla\mathbf{B} \tag{2-15}$$

Using the convective derivative, and since $\partial\mathbf{B}/\partial t = 0$, we can write

$$\nabla\times\mathbf{E} = \mathbf{u}\cdot\nabla\mathbf{B} \tag{2-16}$$

FORCES BETWEEN MOVING CHARGES

Using a vector identity and noting $\nabla \cdot \mathbf{B} = 0$, we obtain

$$\nabla \times \mathbf{E} = -\nabla \times (\mathbf{u} \times \mathbf{B}) \tag{2-17}$$

We can extract $\mathbf{E}$ in the form

$$\mathbf{E} = -\mathbf{u} \times \mathbf{B} - \nabla \psi \tag{2-18}$$

where ψ is a scalar function.

The curl relationship is not unique and we must include an arbitrary gradient. Again here, we know that for $\mathbf{u} = 0$, $\mathbf{E} = -\nabla(q_s/4\pi\varepsilon_0 r)$ and there will be, in general, a velocity-dependent part of ψ to be determined. To do this we first require that $\mathbf{E}$ should be radial and we note that

$$\mathbf{u} = (u\cos\theta)\hat{\mathbf{r}} - (u\sin\theta)\hat{\boldsymbol{\varphi}} \quad \text{and} \quad \mathbf{B} = \frac{q_s u \sin\theta}{4\pi\varepsilon_0 c^2 r^2}\hat{\boldsymbol{\varphi}}$$

So,

$$-\mathbf{u} \times \mathbf{B} = \frac{q_s u^2 \sin^2\theta}{4\pi\varepsilon_0 c^2 r^2}\hat{\mathbf{r}} + \frac{q_s u^2 \sin\theta\cos\theta}{4\pi\varepsilon_0 c^2 r^2}\hat{\boldsymbol{\theta}} \tag{2-19}$$

where $\hat{\mathbf{r}} \times \hat{\boldsymbol{\varphi}} = -\hat{\boldsymbol{\theta}}$ and $\hat{\boldsymbol{\theta}} \times \hat{\boldsymbol{\varphi}} = \hat{\mathbf{r}}$ are used.

We see that ψ must be chosen so that the θ component of ψ will cancel the θ component of $(-\mathbf{u} \times \mathbf{B})$. Then, to determine ψ,

$$\frac{1}{r}\frac{\partial \psi}{\partial \theta} = \frac{q_s u^2 \sin\theta\cos\theta}{4\pi\varepsilon_0 c^2 r^2} \tag{2-20}$$

and the integration over θ gives, by adding a constant of integration $g(r)$,

$$\psi = \frac{q_s u^2 \sin^2\theta}{8\pi\varepsilon_0 c^2 r} + g(r) \tag{2-21}$$

Finally, we choose $g(r)$ to be

$$g(r) = \frac{q_s}{4\pi\varepsilon_0 r} + K\frac{q_s u^2}{4\pi\varepsilon_0 c^2 r} \tag{2-22}$$

Thus, we obtain

$$\psi = \frac{q_s}{4\pi\varepsilon_o r}\left(1 + \frac{1}{2}\frac{u^2}{c^2}\sin^2\theta + K\frac{u^2}{c^2}\right) \quad (2\text{-}23)$$

which again produces the same expression as Eq. (2-13).
We conclude that

$$\mathbf{E} = \frac{q_s}{4\pi\varepsilon_o r^2}\left(1 + \frac{3}{2}\frac{u^2}{c^2}\sin^2\theta + K\frac{u^2}{c^2}\right)\hat{\mathbf{r}} \quad (2\text{-}24)$$

and then,

$$\mathbf{F} = \frac{q_s q_r}{4\pi\varepsilon_o r^2}\left(1 + \frac{3}{2}\frac{u^2}{c^2}\sin^2\theta + K\frac{u^2}{c^2}\right)\hat{\mathbf{r}} \quad (2\text{-}25)$$

We, thus, found a force formula, Eq. (2-25), to describe the forces of two interacting charges in relative motion. As we are looking for, the formula is radial and mutual along the line connecting two charges. This formula, however, includes an unknown constant K. The next step is to determine the constant K. To do this, we apparently need an independent piece of information. As the force between parallel wires carrying neutral currents is generally supposed to be an effect of v^2/c^2, we will examine the possibility of using it to determine K.

The procedure we are going to take here is that by using the newly found formula of Eq. (2-25), we calculate the forces between two current-carrying wires, then, the result is compared with the experimental result we know well. In this way we will be able to determine the unknown constant K. This procedure should be reasonable since the forces between two current carrying wires are well established.

We will assume a particular model for a neutral current and calculate the composite force on q_r by using the force derived in the preceding. First, we will consider the composite force on a stationary charge q_r due to a long straight wire carrying a constant current I (Fig. 2-4). For convenience, we will assume the current to be composed of stationary (with respect to q_r) filament of negative charges forming an infinitely long straight line whose perpendicular distance to q_r is a, and superimposed on the negative filament a similar filament of positive charges moving constantly along the wire with a small velocity **u** relative to q_r. The linear charge density of the negative and positive filaments, λ^- and λ^+, in the same wire will be equal and opposite in sign.

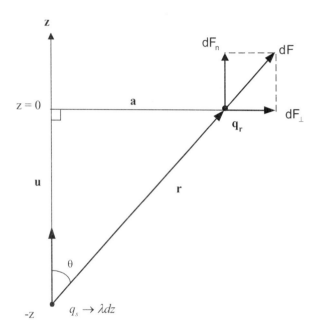

Fig. 2-4

The above assumptions are not what is usually considered rest frame of positive ions. The results are essentially the same. Note here that $\sin\theta = a/r$, $\cos\theta = -z/r$, $\cot\theta = -z/a$ and $dz = a\csc^2 d\theta$.

The combined force on q_r due to the small segment of the current

$$d\mathbf{F} = \left[\frac{q_r \lambda^+ dz}{4\pi\varepsilon_o r^2}\left(1 + \frac{3}{2}\frac{u^2}{c^2}\sin^2\theta + K\frac{u^2}{c^2}\right) + \frac{q_r \lambda^- dz}{4\pi\varepsilon_o r^2}\right]\hat{\mathbf{r}} \qquad (2\text{-}26)$$

And, since $\lambda^- = -\lambda^+$, this reduces to

CHAPTER 2

$$d\mathbf{F} = \frac{q_r \lambda^+ dz}{4\pi\varepsilon_o r^2}\left(\frac{3}{2}\frac{u^2}{c^2}\sin^2\theta + K\frac{u^2}{c^2}\right)\hat{\mathbf{r}} \tag{2-27}$$

Using the relation $dz/r^2 = d\theta/a$, we can express the force in a form suitable for integration over the angle θ. Then,

$$dF_{//} = \frac{q_r \lambda^+ u^2 d\theta}{4\pi\varepsilon_o c^2 a}\left(\frac{3}{2}\sin^2\theta + K\right)\cos\theta \tag{2-28}$$

and

$$dF_\perp = \frac{q_r \lambda^+ u^2 d\vartheta}{4\pi\varepsilon_o c^2 a}\left(\frac{3}{2}\sin^2\theta + K\right)\sin\theta \tag{2-29}$$

Integration over the angle gives

$$F_{//} = 0 \tag{2-30}$$

and

$$F_\perp = \frac{q_r \lambda^+ u^2}{2\pi\varepsilon_o c^2 a}(1+K) \tag{2-31}$$

If we identify $\lambda^+ u$ as I, this can be written as

$$F_\perp = \frac{I q_r u}{2\pi\varepsilon_o c^2 a}(1+K) \tag{2-32}$$

We note that, due to symmetry, $F_{//}$ would be predicted to be zero. Also, and more importantly, $F_\perp$ is clearly due to the relative motion between the "moving" positive charge filament and the stationary charge q_r. In an ordinary electromagnetic discussion, $F_\perp$ must be zero. Unless $K = -1$ in Eq. (2-32), $F_\perp$ cannot be zero. We will see below that $F_\perp$ does not vanish since K will be determined to be $-½$. This is one of the unique results of our new development.

FORCES BETWEEN MOVING CHARGES

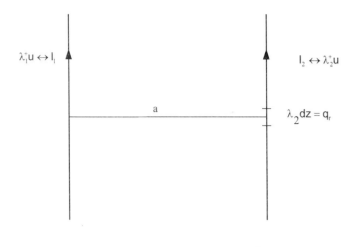

Fig. 2-5

For the case of two long straight neutral currents, I_1 and I_2 shown in Fig. 2-5, we will identify the receiver test charge q_r as $\lambda_2^+ dz$ or $\lambda_2^- dz$, i.e., elements of a stationary negative charged filament parallel to the first current and of a superimposed positively charged filament moving with velocity **u** in the same direction as the first current as shown in Fig. 2-5. We consider both negative charge filaments as stationary.

There are four interactions to consider. The negative filaments have no dynamic interaction; they repel each other by the Coulomb force. The positive filaments have no relative velocity and thus they also have no dynamic interaction, leaving Coulomb repulsion. Each positive filament attracts the other negative filament, with their Coulomb attractions canceling the corresponding repulsive forces enumerated above, leaving the attractive dynamic terms to provide the net force between the two currents.

The dynamic part of the force on $\lambda_2^- dz$ due to λ_1^+ moving at a velocity u ($\lambda_1^+ u = I_1$) is, from Eq. (2-32),

$$dF \text{ (on } 2^- \text{ by } 1^+\text{)} = \frac{I_1 \lambda_2^- u}{2\pi\varepsilon_o c^2 a}(1+K)dz$$

$$dF \text{ (on } 2^- \text{ by } 1^+) = -\frac{I_1 I_2}{2\pi\varepsilon_o c^2 a}(1+K)dz \qquad (2\text{-}33)$$

where $\lambda_2^- = -\lambda_2^+$ and $\lambda_2^+ u = I_2$ are used. By symmetry, the dynamic force on $\lambda_1^+ dz$ due to I_2 is given by

$$dF \text{ (on } 1^+ \text{ by } 2^-) = \frac{I_2 \lambda_1^- u}{2\pi\varepsilon_o c^2 a}(1+K)dz$$

$$dF \text{ (on } 1^+ \text{ by } 2^-) = -\frac{I_1 I_2}{2\pi\varepsilon_o c^2 a}(1+K)dz \qquad (2\text{-}34)$$

where $\lambda_1^- = -\lambda_1^+$ and $\lambda_1^+ u = I_1$ are used.

Thus, the total force per unit length is given by

$$F/L = -\frac{I_1 I_2}{2\pi\varepsilon_o c^2 a}(1+K) \qquad (2\text{-}35)$$

Where the minus sign indicates that the force is attractive and directed from one current to the other perpendicular to the currents.

Now, the well-known result for this force is

$$F/L = -\frac{\mu_o I_1 I_2}{2\pi a} = -\frac{I_1 I_2}{2\pi\varepsilon_o c^2 a} \qquad (2\text{-}36)$$

Equating Eq. (2-35) and Eq. (2-36), we can conclude that

$$K = -\tfrac{1}{2}. \qquad (2\text{-}37)$$

Using this result, we have for the force between two charges, from Eq. (2-25),

$$\mathbf{F} = \frac{q_s q_r}{4\pi\varepsilon_o r^2}\left(1 + \frac{3}{2}\frac{u^2}{c^2}\sin^2\theta - \frac{1}{2}\frac{u^2}{c^2}\right)\hat{\mathbf{r}} \qquad (2\text{-}38)$$

And for the associated "corrected" electric field we have

FORCES BETWEEN MOVING CHARGES

$$\mathbf{E} = \frac{q_s}{4\pi\varepsilon_o r^2}\left(1 + \frac{3}{2}\frac{u^2}{c^2}\sin^2\theta - \frac{1}{2}\frac{u^2}{c^2}\right)\hat{\mathbf{r}} \qquad (2\text{-}39)$$

Equations (2-38) and (2-39) are the results for the interactions of two charges in relative motion with constant velocity. These results are theoretical. Equation (2-38) shows that for interactions of moving charges the Coulomb force is modified by a velocity-dependent term that corresponds to the conventional magnetic effect. In the derivation we "fixed" an arbitrary constant K in order to correctly predict experimentally known forces between two currents. However, experiments to directly verify the force of (2-38) for moving charges are impossibly difficult.

For the problem of anti-parallel currents in long straight wires, we can consider a similar configuration. In this case, the positive filaments will be assumed to be moving in the opposite directions, each with speed u relative to the negative filaments and thus with a relative speed of $2u$. Similar calculations as before give a repulsive force

$$F/L = \frac{\mu_o I_1 I_2}{2\pi a} = \frac{I_1 I_2}{2\pi\varepsilon_o c^2 a} \qquad (2\text{-}40)$$

In this derivation we used Eq. (2-39) with $K = -½$.

These results reinforce our belief that $K = -½$ is a good choice. The coordinate frames used in the above cases are inertial frames and the derived force formula will be valid for every inertial frame. It should also be noted that I_1 and I_2 are defined with respect to the stationary charged filaments, *i.e.*, in the rest frame of the conducting wires.

One of the important findings here is that the force on a "stationary" charge, q_r, due to a straight neutral current is given by, from Eq. (2-32),

$$F_\perp = \frac{I_1 q_r u}{4\pi\varepsilon_o c^2 a} \qquad (2\text{-}41)$$

A neutral current will exert a force on a stationary charge. If the electrons are moving charges as we normally understand for a conductor current, the current will attract a positive charge and repel a negative charge.

We generally understand the force between two current carrying wires by the Lorentz force, *i.e.*, the force exerted on a charge q moving with velocity **v** in a magnetic field **B** is given by

$$\mathbf{F} = q\mathbf{v} \times \mathbf{B}.$$

Where the magnetic field at a distance a away from the current carrying wire is given

$$\mathbf{B} = (\mu_o I / 2\pi a)\,\hat{\phi}.$$

If a charge is stationary with respect to the wire, $\mathbf{v} = 0$ and there should be no Lorentz force. On the other hand, according to the treatment of the preceding argument, the result of Eq. (2-41) gives a non-zero perpendicular force on a stationary charge, because the relative motion between the conductor charges and the stationary charge is not zero and gives rise to a net force. For an application of the perpendicular force to a current-carrying wire, the relative velocity **u** appearing in Eq. (2-41) is considered to be the drift velocity of the electrons of the conducting wire and its magnitude for typical conductors is very small (of the order of 10^{-4} m/s). We should note here that a relativistic consideration could predict a perpendicular force on a stationary charge near a current carrying wire [12].

Electrostatically, a charge near a conductor induces an "image" charge and a first-order Coulomb force arises. Also, it is shown that an electric current segment produces an external electric field proportional to the current [13]. This first order force has been detected and measured [14].

No theoretical prediction will be meaningful, unless an experiment can verify the prediction. If we try an experiment to find the force due to the relative velocity described in Eq. (2-41), an induced charge will appear in the neutral conductor and the resulting Coulomb force will be so dominating that the predicted effect will be undetectable for meaningful verification. A direct observation of this minute additional force seems to be impossible [15,16]. We should realize, however, that it is exactly this force that explains the attraction of parallel currents and the repulsion of anti-parallel currents. So, it is indirectly proven. In the context of our analysis, we have a purely electric interaction. It is rather inconceivable that from the viewpoint of a stationary test charge there would be no electrical difference between a neutral conductor without current and a neutral conductor carrying a current. From the standard viewpoint that is exactly what one expects.

In the following chapter we will discuss the interaction of two charges from the viewpoints by both the above theory and the standard theory including the special theory of relativity. We will be able to clarify the conceptual discrepancy between the two force theories.

Chapter 3. Comparison with relativistic treatment

We now found a force formula, Eq. (2-38), between two moving charges to replace Lorentz's formula. In the new expression the mutual forces are reciprocal, thus satisfying Newton's third law. We have also seen in the derivation that the new formula developed here has, sort of, taken care of the conventional magnetic force of the Lorentz force by adding the "corrected" electrical force to Coulomb's force. So, the new formula takes into consideration a contribution due to v^2/c^2. In addition, the most important difference between this formalism and the conventional one is the view of the velocity **u**, not **v**, the usual velocity of a particle in an inertial frame of reference.

The **v** in the Lorentz force is the velocity of a receiver (test) charge in a particular inertial coordinate system, but the **u** in the present formalism is the relative velocity between interacting charged particles. In the conventional formalism, the magnetic force does not depend only on the velocity of the particle experiencing the magnetic force, but also is different in a moving coordinate system and the force does not simply change sign when the particle labels are exchanged. This dependence on the coordinate system contradicts the assumption of classical mechanics that forces are the same in all inertial coordinate systems. We see a hint here that the theory of relativity is necessary to accommodate the conventional electromagnetism. Einstein tried to fix this problem by his special theory of relativity.

It is a natural turn at this point that we want to compare the present theory with the special theory of relativity. We will, in the following several cases, compare how two interacting charges in relative motion would be different in the two competing theories. We will also solve the paradox (**Case 3** below) encountered at the beginning, one by the special theory of relativity and the other by the new formula.

Case 1 - Two charges moving with the same velocity

Case 1. Let the source charge q_s and receiver (test) charge q_r both be moving with the same velocity relative to a reference frame S. At the time $t = 0$ let the line connecting the instantaneous position of q_s and q_r coincide with the y axis. The situation is shown in Fig. 3-1(a).

CHAPTER 3

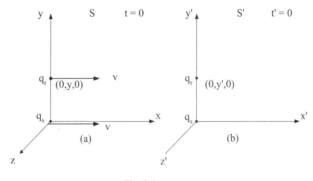

Fig. 3-1

According to the conventional electromagnetic theory, the force on q_r that is moving at a velocity **v** is given by the Lorentz force, $\mathbf{F} = q_r(\mathbf{E} + \mathbf{v} \times \mathbf{B})$. The force is the sum of the electrostatic Coulomb force and the force exerted by a magnetic field on a moving receiver charge. The magnitude of the electric field due to q_s at the position of q_r is $(1/4\pi\varepsilon_0)q_s/y^2$. The magnetic field produced by a moving source charge is given by Biot-Savart's law. The magnitude is $B = (\mu_0/4\pi)q_s v/y^2$ and points to the positive z direction. The force on q_r is only in the y direction and is given by

$$F_y = (1/4\pi\varepsilon_0)q_s q_r / y^2 - (\mu_0/4\pi)v^2 q_s q_r / y^2 \\ = (1-v^2/c^2)k q_s q_r / y^2 \qquad (3\text{-}1)$$

where we used $c^2 = 1/(\varepsilon_0\mu_0)$ and $k = 1/(4\pi\varepsilon_0)$. On the other hand, let us consider the same situation from a moving frame S' that moves at a velocity **v** with respect to S (Fig. 3-1-(b)). Both charges are stationary to each other and the mutual force must be just a Coulomb force, the first term of Eq. (3-1). Which one is correct?

(a) Results using the special theory of relativity

One suggestion would be to treat the problem with application of the special theory of relativity. According to the special theory of relativity, the fields in one inertial frame are related to those of the other frame by relativistic transformations. Suppose that an event observed in a frame of reference S and the same event is observed in another frame of reference S' that is moving at a velocity **v** with respect to S in the direction of the x axis. The

COMPARISON WITH RELATIVISTIC TREATMENT

transformations of the electric and the magnetic field are given in the following form.

Defining $\gamma = (1 - v^2/c^2)^{-1/2}$;

$$E'_x = E_x \quad E'_y = \gamma(E_y - vB_z) \quad E'_z = \gamma(E_z + vB_y)$$
$$B'_x = B_x \quad B'_y = \gamma(B_y + vE_z/c^2) \quad B'_z = \gamma(B_z - vE_y/c^2) \quad (3\text{-}2)$$

Inversely,

$$E_x = E'_x \quad E_y = \gamma(E'_y + vB'_z) \quad E_z = \gamma(E'_z - vB'_y)$$
$$B_x = B'_x \quad B_y = \gamma(B'_y + vE'_z/c^2) \quad B_z = \gamma(B'_z + vE'_y/c^2) \quad (3\text{-}3)$$

Coulomb's law gives the electric force on a test charge, at rest or moving, due to a stationary source charge. We must introduce the Coulomb force in a frame of reference where the source charge is stationary. *S'* of Fig. 3-1(b) is our choice.

In *S'*, the force on q_r is given purely by the Coulomb force. The field at q_r is

$$E'_x = 0 \qquad E'_y = kq_s/y^2 \qquad E'_z = 0 \quad (3\text{-}4)$$

For q_r, with $B' = 0$ in the transformations of Eq. (3-3), we have

$$E_x = 0 \quad E_y = \gamma E'_y \quad E_z = 0$$
$$B_x = 0 \quad B_y = 0 \quad B_z = (v/c^2)\gamma E'_y \quad (3\text{-}5)$$

From (3-5), according to the Lorentz force law, we find for the sum of the electric force and the magnetic force on q_r, *i.e.*, $q_r\mathbf{E}$ and $q_r(\mathbf{v} \times \mathbf{B})$. Since $y' = y$, we have

$$F_y = \gamma kq_s q_r/y^2 - (v^2/c^2)\gamma kq_s q_r/y^2$$
$$= \gamma(1 - v^2/c^2)kq_s q_r/y^2 \quad (3\text{-}6)$$

or
$$F_y = (1 - v^2/c^2)^{1/2} kq_s q_r/y^2 \quad (3\text{-}7)$$

This result is different from Eq. (3-1) by a factor γ. As the velocity v becomes higher, the attractive force term on q_r, the second term of either Eq. (3-1) or Eq. (3-7), increases due to the magnetic effect, and finally the repulsive force is canceled by the attractive force and the total force vanishes at $v = c$.

(b) Result by the new force formula

There is no relative velocity between the two charges. The force involved is only the static Coulomb force. There is no dependence on either velocity, quite different from that by the special theory of relativity.

Case 2 - One charge stationary and the other moving

Case2. It is also interesting to find the second case in which the receiver charge q_r is stationary on the y axis at the point $(0,y,0)$ and only the source charge q_s is moving with the constant velocity $(v,0,0)$ relative to S. At the time $t = 0$, the source charge is located at the origin (Fig. 3-2).

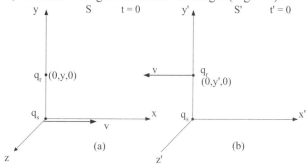

Fig. 3-2

(a) According to the special theory of relativity

In S', q_s is at rest, and therefore we can apply Coulomb's law to obtain the force on q_r. As observed in S' at the time $t' = 0$, q_r is located $(0,y',0)$. The electric field due to q_s at q_r is

$$E_x' = 0 \qquad E_y' = kq_s/y'^2 \qquad E_z' = 0 \qquad (3\text{-}8)$$

Using the transformations we obtain the force on q_r in S. No magnetic force term arises since $\mathbf{v} = 0$ thus $\mathbf{v} \times \mathbf{B} = 0$ for q_r in S. The net force on q_r is

$$F_x = 0 \qquad F_y = \gamma F_y' \qquad F_z = 0 \qquad (3\text{-}9)$$

COMPARISON WITH RELATIVISTIC TREATMENT

With $y' = y$ again, we have

$$F_y = \gamma k q_s q_r / y^2 = \left(1 - v^2/c^2\right)^{-1/2} k q_s q_r / y^2 \qquad (3\text{-}10)$$

In this case, since q_r is at rest no attractive magnetic force arises. As the velocity v approaches to the speed of light, the repulsive force on q_r becomes infinite.

(b) According to the new formula

By our new force formula of Eq. (2-38), the force on q_r will be,

$$F = (1 + v^2/c^2) k q_s q_r / y^2 \approx \gamma^2 k q_s q_r / y^2 \qquad (3\text{-}11)$$

Both results of Eq. (3-11) and Eq. (3-9) show that the force on q_r increases as the source charge moves faster. The difference between them is a factor of γ up to the order of v^2/c^2. We see that the approach of the present development is fundamentally different from that of the special theory of relativity. The difference will become clearer when we consider the next two cases.

Case 3 - Two charges move perpendicular to each other

Case 3. We now resolve the "paradox" cited in the beginning (Fig. 1-1). Consider two charges with velocities at right angles, so that one will cross over the path of the other, but in front of it, so they don't collide. We will alter the situation slightly from the one shown in Fig. 1-1. At some instant ($t = 0$), their relative positions will be shown in S of Fig. 3-3 (a).

(a) Treatment by Lorentz's force and the special theory of relativity

What is the paradox? We will get confused if we consider this case from a non-relativistic viewpoint, that is, by using the Lorentz's force. Let us look at the force on q_r due to q_s and vice versa. On q_s, there is, there is only the electric force from q_r, since q_r makes no magnetic field along its line of motion. On q_r, however, there is again the electric force but, in addition, a magnetic force, since it is moving in a **B**-field made by q_s.

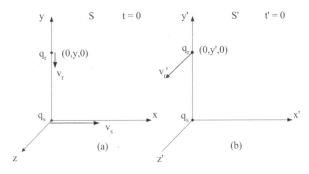

Fig. 3-3

The electric forces on q_s and q_r are equal and opposite. However, there is a sideways (magnetic) force on q_r and no sideways force on q_s.

The force on q_r is

$$F_x = -q_r v_r B_{byqs}, \qquad F_y = q_r E_{byqs} = kq_s q_r/y^2, \qquad F_z = 0. \qquad (3\text{-}12)$$

On the other hand, the force on q_s is

$$F_x = 0, \qquad F_y = q_s E_{byqr} = kq_s q_r/y^2, \qquad F_z = 0. \qquad (3\text{-}13)$$

Action does not equal reaction! The result depends on the coordinate system we choose. This is a paradox. We will get a hint that the theory of relativity is necessary to resolve this confusion.

We will now employ a relativistic method as before. Once more in S', we have

$$E_x' = 0, \qquad E_y' = kq_s/y^2, \qquad E_z' = 0 \qquad (3\text{-}14)$$

Using the transformations given by Eq. (3-3),

$$\begin{aligned} E_x &= 0 & E_y &= \gamma E'_y = \gamma kq_s/y^2 & E_z &= 0 \\ B_x &= 0 & B_y &= 0 & B_z &= \gamma v_s E'_y/c^2 \end{aligned} \qquad (3\text{-}15)$$

Thus, the force on q_r in S is given by the Lorentz force law,

$$F_x = q_r v_r B_z = -\gamma(v_r v_s/c^2) kq_s q_r/y^2, \qquad F_y = \gamma kq_s q_r/y^2 \qquad (3\text{-}16)$$

COMPARISON WITH RELATIVISTIC TREATMENT

Similarly we can compute the force on q_s. The results are illustrated in Fig. 3-4. Compare this with the forces on each charge shown in Fig. 1-1 (b).

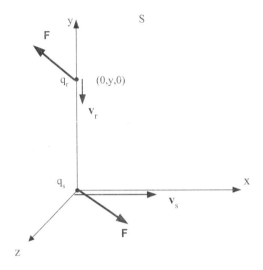

Fig. 3-4

The forces have the same magnitude but the action and reaction do not align along the straight line connecting the charges. There seem to be a torque on the system in S. Does the system rotate around their center of mass by itself? It seems strange. According to the special theory of relativity Newton's third law does not, in general, extend to the relativistic regime. If the two objects are separated in space, the third law is incompatible with the relativity of simultaneity. In Chapter 8, we will discuss a similar situation — the Trouton-Noble experiment — and indicate how a detailed treatment should bail out the trouble.

(b) Treatment by our new theory

On the other hand, our result, Eq. (2-38), gives forces on each charge

$$F = \frac{q_s q_r}{4\pi\varepsilon_o r^2}\left(1 + \frac{v_s^2}{c^2} - \frac{1}{2}\frac{v_r^2}{c^2}\right) \tag{3-17}$$

The directions of the force are opposite to each other and lie along the line connecting two charges. The result satisfies the action-reaction principle, Newton's third law.

In **Cases 1, 2,** and **3** we discussed above, the relativistic results depend upon a choice of a reference frame. The forces that two charges in relative motion experience are different. It is rather puzzling to recognize that the treatment of two charges by the special theory of relativity is not relative as seen in **Case 3**. On the other hand, since the new formulation developed here is based purely on the relative velocity, the results are completely relative; one's view is equal to the other's view. There is no confusion. Especially, **Case 1** is obvious. The force on each charge is simply Coulomb's force since no relative velocity exists — period.

Case 4 - Two charges move on a straight line

Case 4. We need to discuss one more example where two charges are lined on a straight line with a relative velocity. This case is interesting because the new formula gives a *longitudinal* force, but the Lorentz's force does not. How does the special theory of relativity treat this? Does the longitudinal force arise as well? The question will be resolved below. Let the source charge be moving with the constant velocity $(v,0,0)$ relative to S and, at time $t = 0$, be located at the origin. Let the receiver charge q_r be stationary on the x axis at the point $(x,0,0)$. This situation is shown in Fig. 3-5.

(a) According to the special theory of relativity

In S',

$$F'_x = kq_s q_r / x'^2, \qquad F_y = 0, \qquad F_z = 0 \tag{3-18}$$

Transforming back to S,

$$F'_x = \gamma^{-2} kq_s q_r / x^2 = \left(1 - v^2/c^2\right) kq_s q_r / x^2, \quad F_y = 0, \quad F_z = 0 \tag{3-19}$$

The force on q_r has a velocity-dependent term along the line connecting the two charges — a longitudinal force exists.

(b) According to the new formula

By Eq. (2-38), the result is straightforwardly obtained as

$$F = (1 - \tfrac{1}{2} v^2/c^2)\, kq_s q_r/x^2 \approx \gamma^{-1}\, kq_s q_r/x^2, \qquad (3\text{-}20)$$

Once more, we find the difference of a factor γ between the results, Eq. (3-19) and Eq. (3-20).

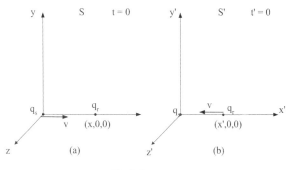

Fig. 3-5

We should note here an important consequence. Suppose that we apply the force of Eq. (3-20) to a neutral current in a conductor. The Coulomb force terms will cancel since equal numbers of the negative and the stationary positive charges exist in the conductor, but the velocity-dependent force parts do not cancel out. This internal force is often advocated as the longitudinal force that uniquely occurs according to Ampere's force law that is closely related with the new force law. In the next Chapter, the so-called Ampere force law will be derived from the new force law of Eq. (2-38) developed in Chapter 2.

Some authors reported observation of such a force [17]. This longitudinal force, however, is not unique to Ampere's force, as we see that the relativistic result of Eq. (3-19) also has a similar velocity-dependent term. Careful measurements should reveal such a force.

We have seen the difference in a factor of γ between the results of the two theories in all the cases above. Which one is correct? We should be able to tell which one is correct by an experiment with such a device as the linear accelerator of Stanford University (SLAC). According to their experiments [18], observations of particles with a variety of velocities have shown that the predictions of the special theory of relativity are real effects. One important difference that was not obvious in the above comparisons is the consequent dynamics of the forces. The special relativity encompasses the relativistic application to the material bodies. The mass of a particle (more correctly its momentum) changes from the rest mass to the relativistic mass in a moving frame and also the perpendicular force is modified by γ but not for the parallel force. The new theory does not require the change of the particle dynamics. The difference of the factor γ compensates the outcomes. In other words, the new theory, without the change of the particle mass, is also able to verify the experimental results by the SLAC.

Chapter 4. Ampere's Force Law

The new force formula developed here is expressed in terms of charged particles. In the middle of the 19th Century, the idea of the elementary particles such as the electron and the proton was not materialized. In these days they talked about more familiar terms, electrical current segments, rather than then-unknown charged particles. From the force formula of Eq. (2-38) we cannot obviously see the force formula between current segments. We therefore deduce the force law for two current segments in the following. The results will lead to rediscover the formula derived in 1823 by Ampere [1775-1836].

Derivation of Ampere's force law

We will model two current segments by two fixed negative charges q_1^- and q_2^- a distant r apart and superimposed moving positive charges q_1^+ and q_2^+. Charge q_1^+ moves with velocity $\mathbf{v}_1$ and q_2^+ moves with $\mathbf{v}_2$ with respect to the negative charges and assume that $v_1 = v_2 = v$ in magnitude. Each q may be written as $\lambda d\ell$ and $\mathbf{v}$; $d\ell$ will specify the direction of motion as appropriate. The geometry is illustrated in Fig. 4-1. We note here that $\varepsilon = \beta - \alpha$, $\gamma + \alpha = \pi$, $\eta + \beta = \pi$, and $q_1^- = -q_1^+$, $q_2^- = -q_2^+$.

As the Coulomb terms add out in the net force, we need only evaluate the dynamic terms. The relative speed of q_1^- and q_2^- is zero, giving no contribution. The relative speed for q_1^+ and q_2^- is v. From Eq. (2-38), and using $q_2^- = -q_2^+$, the force on q_2^- by q_1^+ is

$$dF^{+-} = \frac{q_1^+ q_2^+ v^2}{4\pi\varepsilon_o c^2 r^2}\left(\frac{1}{2} - \frac{3}{2}\sin^2\beta\right) \qquad (4-1)$$

Similarly, for the force on q_1^- by q_2^+ the relative speed is v again and using $q_1^- = -q_1^+$, we have

$$dF^{-+} = \frac{q_1^+ q_2^+ v^2}{4\pi\varepsilon_o c^2 r^2}\left(\frac{1}{2} - \frac{3}{2}\sin^2\gamma\right) \qquad (4-2)$$

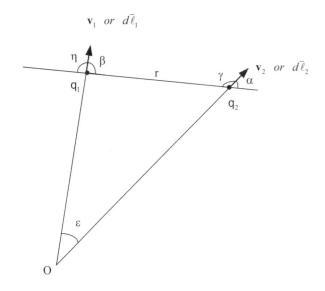

Fig. 4-1

For q_1^+ and q_2^+, their interaction is more complicated. The geometry is indicated in Fig. 4-2. Since $\mathbf{u} = \mathbf{v}_1 - \mathbf{v}_2$, we have drawn $\mathbf{v}_1$ and $-\mathbf{v}_2$ from the same origin. Noting that have the same magnitude, their resultant will bisect the angle $\alpha + \eta$ and have a magnitude of $2v \cos[(\alpha + \eta)/2]$.

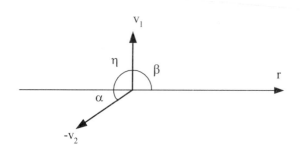

Fig. 4-2 (The magnitudes: $v_1 = v_2 = v$)

The appropriate angle θ will be given by $\theta = (\alpha + \eta)/2 + \beta$. Therefore,

$$dF^{++} = \frac{q_1^+ q_2^+ v^2}{4\pi\varepsilon_o c^2 r^2}\left[\frac{3}{2}\sin^2\left(\frac{\varepsilon+\eta}{2}+\beta\right)-\frac{1}{2}\right]4v\cos^2\left(\frac{\alpha+\mu}{2}\right) \quad (4\text{-}3)$$

Combining the forces, we have

$$dF = \frac{3}{2}\frac{q_1^+ q_2^+ v^2}{4\pi\varepsilon_o c^2 r^2}\left[4\cos^2\left(\frac{\varepsilon+\eta}{2}\right)\sin^2\left(\frac{\alpha+\eta}{2}+\beta\right)-\sin^2\beta-\sin^2\gamma\right]$$
$$-\frac{1}{2}\frac{q_1^+ q_2^+ v^2}{4\pi\varepsilon_o c^2 r^2}\left[4v\cos^2\left(\frac{\alpha+\mu}{2}\right)-2\right] \quad (4\text{-}4)$$

To express this completely in terms of ε, β, and α, we use

$$\gamma = \pi - \alpha, \quad \sin^2\gamma = \sin^2\alpha, \quad \eta = \pi - \beta, \quad \beta = \varepsilon + \alpha$$

So,
$$\eta + \alpha = \pi - \varepsilon,$$
$$\cos(\alpha + \eta) = \sin\varepsilon/2$$

Thus,
$$4\cos^2(\alpha+\eta)/2 - 2 = 2(2\sin^2\varepsilon/2 - 1) = -2\cos\varepsilon$$

Also, since
$$\varepsilon = \beta - \alpha$$
$$4\cos^2\left[(\alpha+\eta)/2\right]\sin^2\left[(a+\eta)/2+\beta\right]-\sin^2\alpha-\sin^2\beta$$

having been written as

$$4\sin^2(\varepsilon/2)\sin^2[(\pi-\varepsilon)/2+\beta]-\sin^2\alpha-\sin^2\beta$$

becomes $\quad \gamma = \pi - \alpha, \quad \sin^2\gamma - \sin^2\alpha, \quad \eta = \pi - \beta, \quad \beta = \varepsilon - \alpha$

$$4\sin^2[(\beta-\alpha)/2]\cos^2[(\alpha+\beta)/2]-\sin^2\alpha-\sin^2\beta$$

The first term is readily reorganized as $(\sin\beta - \sin\alpha)^2$ and we have $-2\sin\alpha\sin\beta$. Thus,

$$dF = \frac{q_1^+ q_2^+ v^2}{4\pi\varepsilon_o c^2 r^2}(\cos\varepsilon - 3\sin\beta\sin\alpha) \tag{4-5}$$

which can be rearranged using

$$\cos\varepsilon = \cos(\beta - \alpha) = \cos\beta\cos\alpha + \sin\beta\sin\alpha$$

so that

$$dF = \frac{q_1^+ q_2^+ v^2}{4\pi\varepsilon_o c^2 r^2}(-2\cos\varepsilon + 3\cos\alpha\cos\beta) \tag{4-6}$$

Finally, writing $q_1^+ \to \lambda_1^+ dl_1$, and $q_2^+ \to \lambda_2^+ dl_2$ and also $\lambda_1^+ v = I_1$, $\lambda_1^+ v = I_2$, we have

$$dF = \frac{I_1 I_2 dl_1 dl_2}{4\pi\varepsilon_o c^2 r^2}(-2\cos\varepsilon + 3\cos\alpha\cos\beta) \tag{4-7}$$

Or, in vector notation, the force law between two current segments is given by

$$d\mathbf{F} = \frac{I_1 I_2}{4\pi\varepsilon_o c^2 r^2}\left[-2(d\mathbf{l}_1 \cdot d\mathbf{l}_2) + 3(d\mathbf{l}_1 \cdot \hat{\mathbf{r}})(d\mathbf{l}_2 \cdot \hat{\mathbf{r}})\right]\hat{\mathbf{r}} \tag{4-8}$$

In either form, with $1/\varepsilon_o = \mu_0 c^2$, this is exactly the force law presented by Ampere in 1823, and the formula is referred to as Ampere's force law. It should be noted that derivation depends on the use of relative velocities, as opposed to individual velocities referred to a convenient frame of reference. The frame of the "stationary" negative filaments (or wires) is the implicit frame for forces. This force is in radial direction only and satisfies Newton's third law.

For comparison, the conventional force that is based on the Biot-Savart law and the Lorentz force can be written as

$$d\mathbf{F} = \frac{\mu_0 I_1 I_2}{4\pi r^2}\left[d\mathbf{l}_2 \times (d\mathbf{l}_1 \times \hat{\mathbf{r}})\right] \tag{4-9}$$

The Ampere's force law and the Biot-Savart-Lorentz (BSL) force are basically similar. As the matter of fact, both forces give the same result when we calculate the force between two closed circuits by integrating over their entire current segments. We cannot distinguish these two forces by considering the interaction between two closed circuits. Then, in all situations do both expressions always agree? One thing we know is, as we showed in Eq. (2-41), that the BSL force cannot predict a force on a stationary charge by a current carrying wire. Another distinction is that the Ampere's force can predict longitudinal forces in the current flow of a conducting channel. The existence of this longitudinal force was stressed by Ampere himself, who devised the so-called Ampere's bridge experiment to show its existence. This longitudinal force has been discussed by Maxwell [19], and a number of authors [20]. Experiments of 1982 published in *Nature* by Graneau [21] suggested such a force in the jet-propulsion in a current flow between liquid and solid conductors and renewed the interest in Ampere's force and stimulated a number of new experiments to detect the unique prediction. The arising force is very small and difficult to differentiate from other effects. But, is this longitudinal force unique to Ampere's force? As we have discussed in the previous section, the longitudinal force should arise in the treatment with the special theory of relativity for the two-charge interaction, Eq. (3-19), as well as the prediction of Ampere's force law or Eq. (3-20). Only a factor of ½ for the term of v^2/c^2 (or a factor γ) makes the difference between them.

Our endeavor to investigate the interaction of two charges in motion with constant velocities resulted in rediscovering the now forgotten formula, Ampere's force law. This formula was once praised by Maxwell and has never been refuted. However, as the concept of fields became standard and we progressed from classical to quantum mechanics, the concepts of energy and momentum that are directly associated with electromagnetic fields (or potentials) became of paramount importance. Ampere's force law has been supplanted by the Biot-Savart law and Lorentz's force law.

We shall further develop the present formalism to higher order corrections in the following section.

Chapter 5. High speed limit and comparison with the special theory of relativity

In Chapter 2, we derived a new force law between moving charges by taking into consideration of the change of the **B** field in relative to the test charge. The dynamic part of the source charge gave the first "correction" to the stationary Coulomb force. In theory, we could continue to seek the correction terms by iteration forever! What will it happen if we continue taking the iteration? When the relative velocity is not very high, the force formula developed in Chapter 2 would be sufficient. For higher relative velocities, however, the first correction may not be sufficient. We need to find the corrections due to higher orders of v^2/c^2.

The new force derived in Chapter 2 implements the source charge's dynamic motion as the first correction to the Coulomb force. In the next section, we will complete two iterations leading to 4th order corrections to Eq. (2-38) of Chapter 2. In the next iteration we will find a familiar coefficient that appears in a binomial expansion of a function. From the expansion, we will try a wild guess to predict the final form of the intended iteration. The results we discover here are turned to be similar to that of the conventional theory including terms by the special theory of relativity. The following development will give formulas that are again different from the conventional ones by a (Lorentz) factor of $\gamma = (1 - v^2/c^2)^{-1/2}$.

Derivation of higher-order term corrections

As in Chapter 2, we still use the notion of the **B** field for convenience in the following development. For higher velocities we should now seek for further corrections by repeating the same procedure. In terms of the "**E** field", we use

$$\mathbf{B} = (1/c^2)(\mathbf{u} \times \mathbf{E}) \tag{5-1}$$

where $\mathbf{u} = u\cos\theta\hat{\mathbf{r}} - u\sin\theta\,\hat{\boldsymbol{\vartheta}}$ and obtain

$$\mathbf{B} = \frac{q_s u^2 \sin^2\theta}{4\pi\varepsilon_o c^2 r^2}\left(1 + \frac{3}{2}\frac{u^2 \sin^2\theta}{c^2} - \frac{1}{2}\frac{u^2}{c^2}\right)\hat{\boldsymbol{\varphi}} \tag{5-2}$$

Now, from Faraday's equation, $\nabla \times \mathbf{E} = -\nabla \times (\mathbf{u} \times \mathbf{B}) - \nabla \times \nabla \psi$, or,

$$\mathbf{E} = -\mathbf{u} \times \mathbf{B} - \nabla \psi \qquad (5\text{-}3)$$

and

$$-\mathbf{u} \times \mathbf{B} = \frac{q_s u^2 \sin^2 \theta}{4\pi\varepsilon_o c^2 r^2}\left(1 + \frac{3}{2}\frac{u^2 \sin^2 \theta}{c^2} - \frac{1}{2}\frac{u^2}{c^2}\right)\hat{\mathbf{r}}$$
$$+ \frac{q_s u^2 \sin\theta \cos\theta}{4\pi\varepsilon_o c^2 r^2}\left(1 + \frac{3}{2}\frac{u^2 \sin^2 \theta}{c^2} - \frac{1}{2}\frac{u^2}{c^2}\right)\hat{\boldsymbol{\theta}} \qquad (5\text{-}4)$$

and $\mathbf{E}$ can be made radial if the θ component of $-\nabla \psi$ is chosen to cancel the θ component of $-\mathbf{u} \times \mathbf{B}$. If we choose

$$\psi = \frac{q_s}{4\pi\varepsilon_o r}\left(1 - \frac{1}{2}\frac{u^2}{c^2} + \frac{1}{2}\frac{u^2 \sin^2 \theta}{c^2} - \frac{1}{8}\frac{u^4}{c^4} + \frac{3}{8}\frac{u \sin^4 \theta}{c^4} - \frac{1}{4}\frac{u^4 \sin^4 \theta}{c^4}\right) \qquad (5\text{-}5)$$

then,

$$-\nabla \psi = \frac{q_s}{4\pi\varepsilon_o r^2}\left(1 - \frac{1}{2}\frac{u^2}{c^2} + \frac{1}{2}\frac{u^2 \sin^2 \theta}{c^2} - \frac{1}{8}\frac{u^4}{c^4} + \frac{3}{8}\frac{u^4 \sin^4 \theta}{c^4} - \frac{1}{4}\frac{u^4 \sin^4 \theta}{c^4}\right)\hat{\mathbf{r}}$$
$$- \frac{q_s}{4\pi\varepsilon_o r^2}\left(\frac{u^2 \sin\theta \cos\theta}{c^2} + \frac{3}{2}\frac{u^4 \sin^3\theta \cos\theta}{c^4} - \frac{1}{2}\frac{u^4 \sin\theta \cos\theta}{c^4}\right) \qquad (5\text{-}6)$$

Clearly, the θ component of Eq. (5-6) will cancel out with that of $-\mathbf{u} \times \mathbf{B}$ of Eq. (5-4).

The first, second, and fourth terms in the r component are not necessary for the cancellation. The first and second terms reproduce the familiar terms in the expression we already have for $\mathbf{F}$. The first term gives the Coulomb force and the second term gives the necessary correction to first order in u^2/c^2 for the correct force between parallel currents. The fourth term is speculative. This forgoing iterative procedure will, if repeated again and again, produce an infinite series leading, when summed, to an expression for a final force expression that is valid for any velocity. Notice that the first, second, and fourth terms in the r component are the first three terms in the expression of

$$\left(1-u^2/c^2\right)^{1/2} \approx 1 - \frac{1}{2}\frac{u^2}{c^2} - \frac{1}{8}\frac{u^4}{c^4} + O\left(\frac{u^6}{c^6}\right)$$

While the first two terms have been justified, it is clear that, at the moment, we have no justification for the remainder of the series. Because $(1 - u^2/c^2)^{1/2}$ is such a ubiquitous factor in electromagnetic and special relativistic analysis, we will assume that we have it, subject to later correction if necessary.

At this point we have, using the relations Eqs. (5-3), (5-4), and (5-6)

$$\mathbf{F} = \frac{q_s q_r \hat{\mathbf{r}}}{4\pi\varepsilon_o r^2} \times \left(1 - \frac{1}{2}\frac{u^2}{c^2} + \frac{3}{2}\frac{u^2\sin^2\theta}{c^2} - \frac{1}{8}\frac{u^4}{c^4} + \frac{15}{8}\frac{u^4\sin^4\theta}{c^4} - \frac{3}{4}\frac{u^4\sin^2\theta}{c^4}\right) \quad (5\text{-}7)$$

If we try a wild speculation without further calculating the higher iterations, we might to say that it is apparent that the repeated iterations will produce

$$\mathbf{F} = \frac{q_s q_r}{4\pi\varepsilon_o r^2} \frac{\left(1-u^2/c^2\right)^{1/2}}{\left(1-\frac{u^2}{c^2}\sin^2\theta\right)^{3/2}} \hat{\mathbf{r}} \quad (5\text{-}8)$$

Let us assume that this is the correct final form. Then an iteration of the force should produce the same expression of the force. We let

$$\mathbf{E} = \frac{q_s}{4\pi\varepsilon_o r^2} \frac{\left(1-u^2/c^2\right)^{1/2}}{\left(1-\frac{u^2}{c^2}\sin^2\theta\right)^{3/2}} \hat{\mathbf{r}} \quad (5\text{-}9)$$

And using $\mathbf{B} = \mathbf{u} \times \mathbf{E}/c^2$, we have

$$\mathbf{B} = \frac{q_s}{4\pi\varepsilon_o r^2} \frac{\left(1-u^2/c^2\right)^{1/2} u\sin\theta}{\left(1-\frac{u^2}{c^2}\sin^2\theta\right)^{3/2}} \hat{\boldsymbol{\varphi}} \quad (5\text{-}10)$$

Then, as we have seen before in Eq. (2-19),

$$-\mathbf{u}\times\mathbf{B} = \frac{q_s q_r}{4\pi\varepsilon_o r^2} \frac{\left(1-u^2/c^2\right)^{1/2}}{\left(1-\frac{u^2}{c^2}\sin^2\theta\right)^{3/2}} \left(\sin\theta\hat{\mathbf{r}} + \cos\theta\hat{\boldsymbol{\varphi}}\right) \quad (5\text{-}11)$$

And following similar to the procedures from Eq. (2-18) to Eq. (2-23), the series for ψ should sum to

$$\psi = \frac{q_s}{4\pi\varepsilon_o r} \frac{\left(1-u^2/c^2\right)^{1/2}}{\left(1-\frac{u^2}{c^2}\sin^2\theta\right)^{1/2}} \quad (5\text{-}12)$$

Thus,

$$-\nabla\psi = \frac{q_s}{4\pi\varepsilon_o r^2} \frac{\left(1-u^2/c^2\right)^{1/2}}{\left(1-\frac{u^2}{c^2}\sin^2\theta\right)^{1/2}} \left[\hat{\mathbf{r}} - \frac{u^2}{c^2}\frac{\sin\theta\cos\theta}{1-\frac{u^2}{c^2}\sin^2\theta}\hat{\boldsymbol{\theta}}\right] \quad (5\text{-}13)$$

so that $\mathbf{E} = -\mathbf{u}\times\mathbf{B} - \nabla\psi$ is induced and we have

$$\mathbf{E} = \frac{q_s}{4\pi\varepsilon_o r^2} \frac{\left(1-u^2/c^2\right)^{1/2}}{\left(1-\frac{u^2}{c^2}\sin^2\theta\right)^{3/2}} \hat{\mathbf{r}} \quad (5\text{-}14)$$

This is our consistent result for the electric field for high velocity cases. For possible use, we note that if we write $\mathbf{B} = \nabla\times\mathbf{A}$, then two potentials are

$$\mathbf{A} = \frac{q_s}{4\pi\varepsilon_o r} \frac{\mathbf{u}}{\left(1-\frac{u^2}{c^2}\right)^{1/2}\left(1-\frac{u^2}{c^2}\sin^2\theta\right)^{1/2}} \quad (5\text{-}15)$$

and

HIGH-SPEED LIMIT

$$\psi = \frac{q_s}{4\pi\varepsilon_o r} \frac{\left(1-u^2/c^2\right)^{1/2}}{\left(1-\frac{u^2}{c^2}\sin^2\theta\right)^{1/2}} \tag{5-16}$$

It is not clear what the significance of these quantities may be, aside from the fact that ψ is necessary in the deduction of the force, **F**.

It is interesting to examine the integral form of Gauss's law, *i.e.*, to calculate the flux, due to a moving source charge, through a closed surface. Using the expression of Eq. (5-14),

$$\oiint \mathbf{E}\cdot\mathbf{n}dA = \int_0^{2\pi}\int_0^{\pi} E_r r^2 \sin\theta\, d\theta\, d\phi$$

$$= \frac{q_s\left(1-\frac{u^2}{c^2}\right)^{1/2}}{2\varepsilon_o}\int_0^{\pi}\frac{\sin\theta\, d\theta}{\left(1-\frac{u^2}{c^2}\sin^2\theta\right)^{3/2}}$$

where we assumed the polar axis parallel to **u**. We obtain

$$\oiint \mathbf{E}\cdot\mathbf{n}dA = \frac{q_s}{\varepsilon_o\sqrt{1-v^2/c^2}} \tag{5-17}$$

The result shows that the total flux is greater than the standard formula by a factor of γ. The electric field we obtained in the above includes the effect of the magnetic field that is treated separately in the conventional treatment. So, we should expect that there should be a difference in the total electric flux for a moving charge from the total electric flux for a stationary charge. In the conventional electromagnetic theory the "pure" electric field is given by

$$\mathbf{E} = \frac{q_s}{4\pi\varepsilon_o r^2}\frac{\left(1-u^2/c^2\right)}{\left(1-\frac{u^2}{c^2}\sin^2\theta\right)^{3/2}}\hat{\mathbf{r}} \tag{5-18}$$

and the Gauss's law for the above gives

$$\oiint \mathbf{E} \cdot \mathbf{n} dA = \frac{q_s}{\varepsilon_0} \qquad (5\text{-}19)$$

which are very similar to the results by the relative motion treatment, but they are different (smaller) by a factor of $\gamma = (1 - u^2/c^2)^{1/2}$. The direct comparison is impossible, since the velocity v is not a relative velocity, but the velocity of the source in some particular reference frame. The electric field on a receiver charge could be given by $\mathbf{F} = q_r \mathbf{E}$ regardless of the movement of the receiver charge. There is an additional force due to the magnetic field of the source charge and the motion of the receiver charge. For a conventional treatment, the special theory of relativity must be used to manipulate these forces and velocities.

Since the force formula of Eq. (5-8) or Eq. (2-38) is one-dimensional, it might be useful for a computer simulation involving the interactions of charged particles (plasmas) for finding magnetic effects or relativistic effects without a more elaborate higher dimensional calculation.

Chapter 6. Scattering and orbit problem

Scattering

Since the force law of Eq. (5-8) in Chapter 5 is a central force and the force takes into consideration of the high-speed effect to the order of (u^2/c^2), it is interesting to seek the effect of the relativistic interaction of two charges that has on the classical Rutherford scattering. The standard treatment using the special theory of relativity is much more complicated. The force law derived here can give a straightforward result.

For two body interactions we assume q_1 to have mass m_1 and q_2 to have mass m_2. The mutual coordinate system will not be an inertial system and so, in order to write the equations of motion we must transform to an inertial frame. The center-of-mass coordinate system will serve our purpose and, for simplicity, in which follows, we will suppose one of the masses to be much larger than the other, *i.e.*, at the center of mass. We let $m_1 >> m_2$ and also $m_2 = m$. The following analysis can be readily modified in the standard way, if desired.

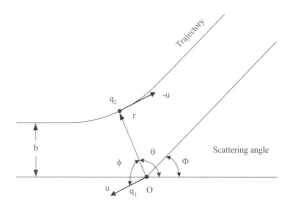

Fig. 6-1

We now write the force using the angle φ to replace θ that was the angle used for the vector **u** and **r** in Chapter 2, as shown in Fig. 6-1. We newly choose a set of coordinates with origin at the center of mass (*i.e.*, at q_1) specified by the polar coordinates, r and θ in the standard way.

Now, $u\cos\varphi$ is the component parallel to **r** and $u\sin\varphi$ is the component perpendicular to **r**. Since $\mathbf{u} = \mathbf{v}_1 - \mathbf{v}_2$ as in Chapter 2 and q_1 is taken to be at rest, $-\mathbf{u}$ is the velocity of q_2 and, in terms of r and θ, the velocity of q_2 can be written as (dr/dt) parallel to **r** and $r(d\theta/dt)$ perpendicular to **r**, thus $u^2\sin^2\varphi = r^2(d\theta/dt)^2$ and $u^2\cos^2\varphi = (dr/dt)^2$, and $u^2 = (dr/dt)^2 + r^2(d\theta/dt)^2 = \dot{r}^2 + r^2\dot{\theta}^2$, where standard dot expressions for the time derivatives are used.

The force is, from Eq. (5-8),

$$\mathbf{F} = \frac{q_1 q_2}{4\pi\varepsilon_o r^2} \frac{\left(1 - u^2/c^2\right)^{1/2}}{\left(1 - \frac{u^2}{c^2}\sin^2\varphi\right)^{3/2}} \hat{\mathbf{r}} \qquad (6\text{-}1)$$

Then, it is written as

$$\mathbf{F} = \frac{q_1 q_2}{4\pi\varepsilon_o r^2} \frac{\left(1 - u^2/c^2\right)^{1/2}}{\left(1 - \frac{r^2\dot{\theta}^2}{c^2}\right)^{3/2}} \hat{\mathbf{r}} \qquad (6\text{-}2)$$

in the center-of-mass frame.

The equations of motion are

$$m\ddot{r} - mr\dot{\theta}^2 = \frac{q_1 q_2}{4\pi\varepsilon_o r^2} \frac{(1 - \frac{\dot{r}^2 + r^2\dot{\theta}^2}{c^2})^{1/2}}{(1 - \frac{r^2\dot{\theta}^2}{c^2})^{3/2}} \hat{\mathbf{r}} \qquad (6\text{-}3)$$

and

$$m\frac{d}{dt}(r^2\theta) = 0 \qquad (6\text{-}4)$$

where we have used the fact that central force motion lies in a plane determined by the initial velocity and the force vector. A first integral of the motion is obtained by integrating the second of the above equations and expressing the conservation of angular momentum,

$$mr^2\dot{\theta} = L \qquad (6\text{-}5)$$

and let the first integral be

$$r^2\dot{\theta} = \ell = \frac{L}{m} \qquad (6\text{-}6)$$

where ℓ is the angular momentum per unit mass. We use the first integral to rewrite the force and obtain

$$\ddot{r} - r\dot{\theta}^2 = \frac{q_1 q_2}{4\pi\varepsilon_o mr^2} \frac{\left(1 - \frac{\dot{r}^2}{c^2} - \frac{\ell^2}{r^2 c^2}\right)^{1/2}}{\left(1 - \frac{\ell^2}{r^2 c^2}\right)^{3/2}} \qquad (6\text{-}7)$$

It is convenient to set $K = q_1 q_2 / 4\pi\varepsilon_o m$ and write the remaining equation (after completely eliminating θ) as

$$\ddot{r} - \frac{\ell^2}{r^3} = \frac{K}{r^2} \frac{\left(1 - \frac{\dot{r}^2}{c^2} - \frac{\ell^2}{r^2 c^2}\right)^{1/2}}{\left(1 - \frac{\ell^2}{r^2 c^2}\right)^{3/2}} \qquad (6\text{-}8)$$

Now, we make an analysis of a certain class of central, velocity-dependent forces and orbit problem.

We will discuss scattering problems using the above equations and derive a higher order correction term arising from the velocity-dependent part of the force to the Rutherford scattering cross section. More general orbit problems using Weber's force will be discussed in Chapter 7.

To write the velocity-dependent central force equations, we make a change of variable. Let $z = \ell/rc$ where c is the speed of light. For the scattering, the initial condition occurs when $z \to 0$ and $r \to \infty$,

$$\frac{d^2z}{d\theta^2}+z=-\frac{KA}{\ell c}\frac{1}{\left(1-z^2\right)^{3/2}}-\frac{K^2}{\ell^2 z^2}\frac{z}{\left(1-z^2\right)^2} \qquad (6\text{-}9)$$

where $A = (1 - u_o^2/c^2)^{1/2}$ with u_o being the initial speed of the scattering particle. We define the impact parameter by b by $\ell = bu_0$.

In order to get an approximate solution, we used a perturbation technique on this second order equation. In doing above, we expect $z \ll 1$ since $z = \ell/rc = (b/r)(u_o/c)$ and $r > b$ and $u \ll c$ for the purpose of deriving the scattering angle. Expanding the right hand side of the equation in powers of z and keeping terms up to z^2 we obtain

$$\frac{d^2z}{d\theta^2}+\left(1+\frac{K^2}{\ell^2 c^2}\right)z=-\frac{KA}{\ell c}-\frac{3}{2}\frac{KA}{\ell c}z^2 \qquad (6\text{-}10)$$

After a several lines of calculations (see Appendix), we obtain, for the scattering cross section in terms of the impact parameter and the scattering angle Φ,

$$\sigma(\Phi)=\frac{1}{4}\frac{K^2}{u_o^4}\csc^4\frac{\Phi}{2}\left(1+\frac{u_o^2}{c^2}\sin\frac{\Phi}{2}\right) \qquad (6\text{-}11)$$

Clearly, we have a first order speed correction to the familiar low speed limit. An interesting point here is that the correction term depends on the sign of the angle Φ.

This dependency on the attractive and repulsive forces is not observed in the classical Rutherford scattering treated by Maxwell's electrodynamics. Weber's theory that we will discuss in the following chapter gives a more general result for the scattering angle [22].

Orbit Problem

We further study the central force problem to seek the relativistic effect on the constant of motion. We rewrite the equation of motion, Eq. (6-7), as

SCATTERING

$$m\frac{d\mathbf{u}}{dt}\cdot\hat{\mathbf{r}} = \frac{K}{r^2}\frac{\left(1-\frac{u^2}{c^2}\right)^{1/2}}{\left(1-\frac{\ell^2}{r^2c^2}\right)^{3/2}} \quad (6\text{-}12)$$

where $\mathbf{u} = d\mathbf{r}/dt$ and $\hat{\mathbf{r}}$ is the radial unit vector. Rewriting Eq. (6-12),

$$\frac{m}{\left(1-\frac{u^2}{c^2}\right)^{1/2}}\frac{d\mathbf{u}}{dt}\cdot\hat{\mathbf{r}} = \frac{K}{r^2}\frac{1}{\left(1-\frac{\ell^2}{r^2c^2}\right)^{3/2}} \quad (6\text{-}13)$$

Having a dot product with $\mathbf{u} = d\mathbf{r}/dt$ and performing integration, we have

$$mc^2\frac{d}{dt}\left(\sqrt{1-\frac{u^2}{c^2}}\right) = K\frac{d}{dt}\left(\frac{1}{r\sqrt{1-\frac{\ell^2}{r^2c^2}}}\right) \quad (6\text{-}14)$$

Thus, we obtain

$$-mc^2\sqrt{1-\frac{u^2}{c^2}} + \frac{K}{r}\frac{1}{\sqrt{1-\frac{\ell^2}{r^2c^2}}} = A \quad (6\text{-}15)$$

where A is an integration constant.

If we assume the second term is the potential and let $A - mc^2 = E$, a new constant, we have for the constant of integration

$$mc^2\left(1-\sqrt{1-\frac{u^2}{c^2}}\right) + \frac{K}{r}\frac{1}{\left(1-\ell^2/r^2c^2\right)^{1/2}} = E \quad (6\text{-}16)$$

Suppose that $u = 0$ at $r = \infty$ then $E = 0$. In a particular case where $q_1 = +e$ of the proton and $q_2 = -e$ of the electron and for a radial motion, *i.e.*, a head-on interaction, $\ell = 0$. When $u = c$, Eq. (6-16) becomes

$$mc^2 = \frac{e^2}{4\pi\varepsilon_o} \frac{1}{r_e} \qquad (6\text{-}17)$$

We have the closest approach $r_e = e^2/4\pi\varepsilon_o mc^2$, the classical radius of the electron. For a limit that $u \ll c$, this equation becomes

$$\frac{1}{2}mu^2 + \frac{K}{r}\frac{1}{\left(1-\ell^2/r^2c^2\right)^{1/2}} = E \qquad (6\text{-}18)$$

If the particle moves at the speed of light, the particle could be a photon. Then, the interaction potential would be zero. From Eq. (6-16), we have

$$mc^2 = E \qquad (6\text{-}19)$$

The photon appears to possess a mass m. Since the mass of a photon should be zero, the quantity m should be considered an apparent mass for a photon that possesses an energy E.

Chapter 7. A more general theory — an encounter with Weber's force law

In the previous derivation of Ampere's force law, we considered that the charges are moving at constant velocities. The velocity term of the moving charge gave rise to a correction to the static Coulomb force. The correction corresponds to the magnetic force in the standard theory. So, if we now include the rate of velocity change, *i.e.*, the relative acceleration of charges, what will the result be? We can expect that the result should result in a higher order correction to the electric force due to the acceleration, possibly similar to Faraday's law. Faraday's law states that the changing magnetic flux through a circuit induces an *emf* in the circuit, that is, an electric field is produced by the changing magnetic field. We will next discuss the two charges in relative motion including their accelerations in their relative coordinates and show that the result will lead to Weber's force law now forgotten in the development of the present-day electromagnetic theory.

We will derive the Weber force using Maxwell's term to Ampere's law, that is, the time derivative of an electric field gives rise to a magnetic field. As in Chapter 2, consider a source charge q_s located at $\mathbf{r}_s$ and a receiver test charge q_r located at $\mathbf{r}_r$ with velocities $\mathbf{v}_s$ and $\mathbf{v}_r$, and accelerations $\mathbf{a}_s$ and $\mathbf{a}_r$, respectively. We formulate our calculation in terms of the relative position, $\mathbf{r} = \mathbf{r}_r - \mathbf{r}_s$, the relative velocity $\mathbf{u} = \mathbf{v}_r - \mathbf{v}_s$, and relative acceleration $\mathbf{a} = \mathbf{a}_r - \mathbf{a}_s$, of the receiver charge to the source charge. Working primarily with the two curl equations that are a pair of coupled partial differential equations, we seek an iterative solution for the electric field at the location of the receiver charge. In the following, we note that subscripts 0 and 1 for $\mathbf{E}$ and $\mathbf{B}$ refer to the zeroth order and first order quantities respectively. The initial assumption is that Coulomb's law gives the instantaneous value of the field due to the source at the receiver

$$\mathbf{E}_o = \frac{q_s}{4\pi\varepsilon_o}\mathbf{r}, \qquad (7\text{-}1)$$

which is not an explicit function of the time or any of the derivatives of the relative displacement such as the velocity and the acceleration. We imagine the source charge and its field moving relatively to the receiver so that the field vector at the receiver has a time dependence that is due to the relative motion. This will allow us to calculate an expression for the auxiliary vector, $\mathbf{B}_o$, as follows

$$\nabla \times \mathbf{B}_o = \frac{1}{c^2} \frac{d\mathbf{E}_o}{dt} \tag{7-2}$$

Note that Eq. (7-2) is correctly written in terms of a total time derivative [23]. The familiar version in terms of a partial derivative assumes an explicit time variation of a stationary field. Since, in general, the electric field may be a function of the time, the relative displacement between the source and the receiver, and various time derivatives of the displacement, its total time derivative is given by

$$\frac{d\mathbf{E}}{dt} = \frac{\partial \mathbf{E}}{\partial t} + (\mathbf{u} \cdot \nabla)\mathbf{E} + (\mathbf{a} \cdot \nabla_u)\mathbf{E} + \cdots \tag{7-3}$$

and we have $\nabla \times \mathbf{B}_o = (\mathbf{u} \cdot \nabla)\mathbf{E}_o(\mathbf{r})/c^2$.

Since $\nabla \cdot \mathbf{u} = 0$, and $\nabla \cdot \mathbf{E}_o = 0$ (everywhere except at $\mathbf{r} = 0$), we can write the last equation as

$$\nabla \times \mathbf{B}_o = -\frac{1}{c^2} \nabla \times (\mathbf{u} \times \mathbf{E}_o)$$

that to within the gradient of an arbitrary function we have

$$\mathbf{B}_o = -\frac{1}{c^2} \mathbf{u} \times \mathbf{E}_o$$

and we note that when $\mathbf{v}_r = 0$, this reduces to the Biot-Savart expression,

$$\mathbf{B}_o = -\frac{1}{c^2} \mathbf{v}_s \times \mathbf{E}_o$$

The next step is to use the expression for $\mathbf{E}_o$ to calculate an iterated formula for the electric field $\mathbf{E}_1$ from the other Maxwell's curl equation

$$\nabla \times \mathbf{E}_1 = -\frac{d\mathbf{B}_o}{dt} \tag{7-3}$$

As before

$$\frac{d\mathbf{B}_o}{dt} = \frac{\partial \mathbf{B}_o}{\partial t} + (\mathbf{u}\cdot\nabla)\mathbf{B}_o + (\mathbf{a}\cdot\nabla_u)\mathbf{B}_o + \cdots$$

Now $\mathbf{B}_o$ is an explicit function of $\mathbf{r}$ and $\mathbf{v}$, so that

$$\nabla\times\mathbf{E}_1 = \frac{1}{c^2}[(\mathbf{u}\cdot\nabla)(\mathbf{u}\times\mathbf{E}_o) + (\mathbf{a}\cdot\nabla_u)(\mathbf{u}\times\mathbf{E}_o)] \tag{7-4}$$

Since $\nabla\mathbf{u} = 0$, $\nabla_u\mathbf{u} = \mathbf{I}$ (the identity dyadic), and $\nabla_u\mathbf{E}_o = 0$, we find

$$\nabla\times\mathbf{E}_1 = \frac{1}{c^2}\left[\mathbf{u}\times(\mathbf{u}\cdot\nabla)\mathbf{E}_o + \mathbf{a}\times\mathbf{E}_o\right] \tag{7-5}$$

If we write $\mathbf{E}_o = -\nabla\phi_o$, then after considerable rearrangement, we can display the right-hand side of Eq. (7-5) as a curl

$$\nabla\times\mathbf{E}_1 = \frac{1}{c^2}\left[\nabla\times\phi_o\mathbf{a} - \frac{3}{2}\nabla\times\frac{(\mathbf{u}\cdot\mathbf{r})^2}{r^2}\mathbf{E}_o\right] \tag{7-6}$$

hence

$$\mathbf{E}_1 = \frac{\phi_o\mathbf{a}}{c^2} - \frac{3}{2}\frac{(\mathbf{u}\cdot\mathbf{r})^2}{c^2 r^2}\mathbf{E}_o - \nabla\phi_1$$

In the static limit when $\mathbf{a}\to 0$ and $\mathbf{u}\to 0$, we must have $\mathbf{E}_1\to 0$ and so we can write $\phi_1 = \phi_0 + \phi_2$ provided with $\phi_2 \to 0$ in the above case, Then

$$\mathbf{E}_1 = \mathbf{E}_o + \frac{\phi_o\mathbf{a}}{c^2} - \frac{3}{2}\frac{(\mathbf{u}\cdot\mathbf{r})^2}{c^2 r^2}\mathbf{E}_o - \nabla\phi_2 \tag{7-7}$$

It is an interesting fact that $\mathbf{E}_1$ can be made to be purely radial vector by choosing $\phi_2 = (\mathbf{a}\cdot\mathbf{r})\phi_o/c^2 + \phi_3$, with ϕ_3 tending to zero in the static limit.

We then have

$$\mathbf{E}_1 = \mathbf{E}_o + \frac{(\mathbf{a}\cdot\mathbf{r})}{c^2}\mathbf{E}_o - \frac{3}{2}\frac{(\mathbf{u}\cdot\mathbf{r})^2}{c^2 r^2}\mathbf{E}_o - \nabla\phi_3 \tag{7-8}$$

We note that this can be done, because not only is this the simplest choice vectorially, it is also the most rational choice from a physical point of view. Even more satisfying is the fact that this expression can be written solely in terms of the relative separation and its time derivatives. We note that

$$u^2 = \frac{(\mathbf{u}\cdot\mathbf{r})^2}{r^2} + \frac{|\mathbf{u}\times\mathbf{r}|^2}{r^2}$$

and the radial component of the acceleration

$$\mathbf{a}\cdot\hat{\mathbf{r}} = \ddot{r} - \frac{|\mathbf{u}\times\mathbf{r}|^2}{r^2}$$

so that

$$\mathbf{a}\cdot\mathbf{r} = r\ddot{r} - u^2 + \frac{(\mathbf{u}\cdot\mathbf{r})^2}{r^2}$$

and, since $\mathbf{u}\cdot\hat{\mathbf{r}} = \dot{r}$, we find

$$\mathbf{E}_1 = \mathbf{E}_o\left(1 + \frac{r\ddot{r}}{c^2} - \frac{\dot{r}^2}{2c^2}\right) - \frac{u^2}{c^2}\mathbf{E}_o - \nabla\phi_3 \qquad (7\text{-}9)$$

and if we choose $\phi_3 = u^2\,\phi_o/c^2 + \phi_4$, then

$$\mathbf{E}_1 = \mathbf{E}_o\left(1 + \frac{r\ddot{r}}{c^2} - \frac{\dot{r}^2}{2c^2}\right) - \nabla\phi_4 \qquad (7\text{-}10)$$

or, to within an arbitrary gradient, we can write

$$\mathbf{E}_1 = \mathbf{E}_o\left(1 + \frac{r\ddot{r}}{c^2} - \frac{\dot{r}^2}{2c^2}\right) \qquad (7\text{-}11)$$

The force is obtained by multiplying this expression by the receiver test charge:

$$\mathbf{F} = \frac{q_s q_r}{4\pi\varepsilon_o r^2}\left(1 + \frac{r\ddot{r}}{c^2} - \frac{\dot{r}^2}{2c^2}\right)\hat{\mathbf{r}} \qquad (7\text{-}12)$$

This expression of force is precisely the force advocated by Weber. It should be emphasized that the force does not depend on the relative velocity or on the relative acceleration, rather that it depends only on the separation and its first and second time derivatives.

Weber's law

We have shown above that starting from Maxwell's equations we derived Weber's force. Historically, Weber's theory and Maxwell's theory were two competing formulations of electromagnetic problems. It is clear that Weber's force has a significant degree of compatibility. We will show later the Weber force can give rise to the familiar expressions for Faraday's law, vector potential, and mutual inductance expressed in relative coordinates.

Including the relative acceleration for the interaction of two moving charges, we now encounter the force derived by Weber in 1846. Weber did not use Maxwell's formula to derive his force law. He independently derived his formula purely based on his experimental observations. Wilhelm Edward Weber was a German experimental physicist born in 1804. He was attracted to electromagnetic problems, a subject on which he had never worked before, when he met C. F. Gauss in 1828, who became his mentor and a collaborator. Historically in 1785, Coulomb's law was given for charges e and e' in electrostatic units separated by a distance r as

$$F = \frac{ee'}{r^2} \qquad (7\text{-}13)$$

Repulsion or attraction occurs according as this expression has a positive or negative value. In 1823 Ampere obtained his force between the current elements ids and ids' when they are separated by a distance r, that is written as

$$d^2 F = -\frac{ii'dsds'}{r^2}\left(\cos\varepsilon - \frac{3}{2}\cos\theta\cos\theta'\right) \qquad (7\text{-}14)$$

Again, repulsion or attraction occurs as this expression has a positive or negative value, and ε is the angle between the positive direction of the currents in ds and ds', and θ and θ' are the angles between these positive directions and the connecting right line between them. Weber, in an attempt to unify electrostatics (Coulomb force) with electrodynamics (Ampere's force), proposed in 1848 that each current element in metallic conductors should be considered as charges in motion. He assumed that the current in metallic conductors is due to an equal amount of positive and negative charges moving in opposite directions relative to the wire with equal velocities. With these ideas in 1846, he arrived at his formula for the force between two charges in relative motion as

$$F = \frac{ee'}{r^2}\left[1 - \frac{1}{c_w^2}\left(\frac{dr}{dt}\right)^2 + \frac{2}{c_w^2} r \frac{d^2r}{dt^2}\right] \qquad (7\text{-}15)$$

The mathematical procedure of the derivation in details can be found in [24, 25]. The constant c_w, which appears in this expression, is the ratio between the electrodynamic and electrostatic units of charge. It has the dimension of a velocity (m/s). If we let $c_w = \sqrt{2}c$ where c is the speed of light, this formula becomes that of the present-day expression of Eq. (7-12).

The first measurement of the ratio between electro*dynamic* and electrostatic units of charge, c_w, was performed in 1856 and found to be 4.39×10^8 m/s [26], that means $\sqrt{2}$ times of c (the speed of light), known then as the ratio of electro*magnetic* and electrostatic unit of charge. In 1857, using Weber's formula, the equation of the conservation of charge, and from the generalized Ohm's law, Weber and Kirchhoff independently predicted the existence in a conducting circuit of negligible resistance of periodic modes of oscillation of the electric current whose velocity of propagation had the value of c, the speed of light. They were the first to derive the wave equation (the telegraph equation) describing a signal in the current propagating along a wire, that is,

$$\frac{\partial^2 I}{\partial s^2} - \frac{1}{c^2}\frac{\partial^2 I}{\partial t^2} = K \frac{\partial I}{\partial t} \qquad (7\text{-}16)$$

where I is the current, s is the distance along the wire from a fixed origin and K is a constant proportional to the resistivity of the wire. This result was independent of the cross section of the wire, of its conductivity, and of the density of electricity in the wire [27]. We note here that they did not utilize the concept of ether, of displacement current, nor of retarded time. Maxwell introduced the term in the displacement current in the circuital law for the magnetic field, the term with c^2 in Eq. (2-1), in 1861. He obtained that an electromagnetic signal would propagate in the ether with a velocity $c = (\varepsilon_o \mu_o)^{-1/2}$.

In 1889 the British physicist Oliver Heaviside [1850-1925] obtained the expression of the force for a charge moving in a magnetic field [28],

$$\mathbf{F} = q\mathbf{v} \times \mathbf{B} \qquad (7\text{-}17)$$

This form is what we accept today for the magnetic force acting on a charge q. Later Lorentz presented his force law in 1895 of the form

$$\mathbf{F} = q(\mathbf{E} + \mathbf{v} \times \mathbf{B}) \qquad (7\text{-}18)$$

Isn't this force basically the same as Heaviside's expression? It is the one just adding an extra force due to an electric field. A very important question is the meaning of the velocity $\mathbf{v}$. Is the velocity of q relative to what? According to historical document [28], Heaviside meant the velocity of q relative to the medium through which is moving — the medium whose magnetic permeability is μ. The Lorentz's $\mathbf{v}$ of Eq. (7-18) had the same meaning when first envisioned. If the medium is vacuum, to Lorentz's, it is the ether that is in a state of absolute rest relative to the frame of the universe. The significant change for the velocity that appears in Lorentz's force came with Einstein's paper of 1905 on the special theory of relativity. In it Einstein begins to interpret this velocity of charge q relative to an observer or a frame of reference. Since then, the standard accepted formula for the force on a charge q in electromagnetism is given by the Lorentz's force in which the velocity is interpreted as the velocity relative to an observer's frame of reference. Lorentz's force does not satisfy Newton's third law. We say that the electric and magnetic fields are observer-dependent and that $\mathbf{E}$ transforms into $\mathbf{B}$ depending on the frame of reference. In Chapter 3 we discussed some of examples involving the relativistic transformations the fields based on the special theory of relativity.

Weber's force on the other hand describes the force between two charges in motion strictly from a relative perspective. The velocity is the first time derivative and the acceleration is the second time derivative of the separation, and the resulting force formula does need an introduction of the extra conceptual field — the magnetic field. The Weber law consequently satisfies Newton's third law of action-reaction principle. Prior to the widespread acceptance of Maxwell's electromagnetic theory with Lorentz's force, the formulation of electrodynamics by Weber was quite successful in correctly describing the known phenomena of electromagnetism. This was acknowledged by Maxwell [29] whose argument against Weber's force was based primarily on conceptual grounds.

Compatibility with Maxwell's theory

We will now show that the Weber force can give rise to the familiar expressions for Faraday's law, vector potential and mutual inductance when these are expressed in relative coordinates [4, 30].

Consider a fixed loop of receiver charge. When integrated over the closed loop, the scalar product of the force per unit charge with a vector element of the loop is, by definition, the *emf* produced in the loop by the force. Thus we have, using Eq. (7-11),

$$emf = \oint_R d\bar{\ell}_R \cdot \mathbf{E}_0 \left(1 + \frac{r\ddot{r}}{c^2} - \frac{\dot{r}^2}{2c^2}\right) \tag{7-19}$$

For any scalar function, we have

$$\int_R d\bar{\ell}_R \cdot \nabla f = \int_R df = 0$$

so that

$$emf = \int_R d\bar{\ell}_R \cdot \left[\mathbf{E}_o \left(1 + \frac{r\ddot{r}}{c^2} - \frac{\dot{r}^2}{2c^2}\right) - \frac{\nabla \phi_4}{c^2}\right]$$

and, using Eq. (7-1), the integrand can be expressed as

$$emf = \int_R d\ell_R \cdot \left[\frac{\phi_o \mathbf{a}}{c_2} - \frac{3}{2c^2}\frac{(\mathbf{u} \cdot \mathbf{r})^2}{r^2}\mathbf{E}_o - \nabla \phi_1\right]$$

Using Stokes's theorem, we have

$$emf = \iint_\sigma d\sigma \mathbf{n} \cdot \nabla \times \left[\frac{\phi_o \mathbf{a}}{c^2} - \frac{3}{2c^2} \frac{(\mathbf{u}\cdot\mathbf{r})^2}{r^2} \mathbf{E}_o \right]$$

and R bounds σ. Now, by reversing the development leading from Eq. (7-3) to Eq. (7-6), we find

$$emf = \iint_\sigma d\sigma \mathbf{n} \cdot \frac{d\mathbf{B}_o}{dt}$$

and since the receiver loop is fixed, we can write

$$emf = -\frac{d}{dt}\phi_B \qquad (7\text{-}20)$$

where

$$\phi_B = \iint_\sigma d\sigma \mathbf{n} \cdot \mathbf{B}_o$$

and when $\mathbf{v}_r = 0$, this reduces to the usual definition of magnetic flux and Eq. (7-19) becomes the familiar statement of Faraday's law of induction.

If we define the auxiliary vector $\mathbf{A}$ by

$$\mathbf{A} = -\frac{\phi_o \mathbf{u}}{c^2} + \nabla g \qquad (7\text{-}21)$$

with g an arbitrary scalar, then

$$\mathbf{B}_o = -\frac{1}{c^2}\mathbf{u}\times\mathbf{E}_o = \nabla \times \mathbf{A}$$

and we write

$$emf = -\frac{d}{dt}\iint_\sigma d\sigma \mathbf{n} \cdot \nabla \times \mathbf{A} = -\frac{d}{dt}\oint_R d\bar{\ell}_R \cdot \left(-\frac{q_s \mathbf{u}}{4\pi\varepsilon_o c^2 r} \right)$$

Next, we envision a closed loop of S of source charge moving collectively to form a current defined by $I\,d\bar{\ell}_s = -q_s\mathbf{u}$. Integrating over the source loop and using relation $\mu_0 = 1/\varepsilon_0 c^2$, we obtain

$$emf = -\frac{d}{dt}\oint_R d\bar{\ell}_R \cdot \oint_S \frac{\mu_o I}{4\pi r}d\bar{\ell}_S$$

which is valid for any time variation of the source. In the particular case of two fixed loops, this is usually written as

$$emf = -\frac{\mu_o}{4\pi}\oint_R \oint_S \frac{d\bar{\ell}_R \cdot d\bar{\ell}_S}{r}\frac{dI}{dt} \qquad (7\text{-}22)$$

or

$$emf = -M\frac{dI}{dt} \qquad (7\text{-}23)$$

where M is the mutual inductance defined to be

$$M = \frac{\mu_o}{4\pi}\oint_R \oint_S \frac{d\bar{\ell}_R \cdot d\bar{\ell}_S}{r} \qquad (7\text{-}24)$$

It is clear that a significant degree of compatibility exists between Weber's force and Maxwell's equation.

In 1848, two years after he published his force law, he presented the velocity-dependent potential energy associated with Weber's force. The potential energy is written as

$$U = \frac{q_s q_r}{4\pi\varepsilon_o}\frac{1}{r}\left(1 - \frac{\dot{r}^2}{2c^2}\right) \qquad (7\text{-}25)$$

We can derive Weber's force from

$$\mathbf{F} = -\left(\frac{dU}{dr}\right)\hat{\mathbf{r}} \qquad (7\text{-}26)$$

Weber's potential energy was the first example in physics of a potential energy that depends not only on the distance between the interacting bodies, but also on their time derivatives (velocities). We note here that the associated Lagrangian energy is slightly different. The Lagrangian energy is given by

$$U_L = \frac{q_s q_r}{4\pi\varepsilon_o} \frac{1}{r}\left(1 + \frac{\dot{r}^2}{2c^2}\right) \quad (7\text{-}27)$$

with the Lagrangian $L = K - U_L$ in which $K = \frac{1}{2} m_s v_s^2 + \frac{1}{2} m_r v_r^2$.

The Hamiltonian H for a two-body system is given by

$$H = \left[\sum_1^6 \dot{q}_k \frac{\partial L}{\partial \dot{q}_k}\right] - L \quad (7\text{-}28)$$

In Weber's electrodynamics, the Hamiltonian is the conserved energy of the system, namely

$$H = E = K + U \quad (7\text{-}29)$$

where U is of Eq. (7-25) and not U_L. This also happens in the standard classical electromagnetism.

Chapter 8. Applications of Weber's electrodynamics

We will now elucidate the unique characteristics of Weber's electrodynamics by discussing some of electromagnetic problems.

When we try to solve electromagnetic problems, we some times encounter a situation that gives one answer when analyzed one way, and a different answer when analyzed another way — a paradox. We will show, in the following, some of the well-known paradoxes and discuss how Weber's theory will account for these apparent paradoxes.

Trouton-Noble experiment

The first example is the famous Trouton-Noble experiment [31]. A charged parallel-plate capacitor moves with a velocity **v** with a slant angle with its surface vector, as illustrated in 8-1. Trouton and Noble analyzed this situation by considering minimization of magnetic energy arising from the motion of the electric field of the capacitor and obtained a torque on the capacitor with the magnitude proportional to the speed. The original Trouton-Noble experiment was performed for the detection of "ether" and it yielded a null result. The Trouton-Noble experiment is often regarded as the electrostatic equivalent of the Michelson-Morley experiment. People believed that the Trouton-Noble experiment contradicted classical physics and lent support to Einstein's special theory of relativity. Hayden, in 1993, repeated the same experiment with 10^5 times greater sensitivity, and again obtained a null result [32]. However, he also showed that the experiment did not contradict either classical theory or Einstein's theory. The nature gives only the correct answer and that is right way. So, this problem should not be called as a paradox. Only confusion in our understanding has drawn much interest in the past. The correct explanation is a complicated one. We must take into consideration of the energy flow associated with electric and magnetic fields in the surrounding space. We will omit the detailed result here [33, 34]. The analysis by Weber's theory, however, is obvious. The relative velocities between the charged plates are zero. No torque should arise.

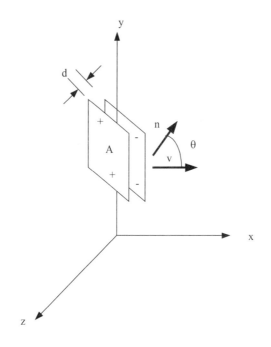

Fig. 8-1. Trouton-Noble experiment

Rotating solenoid

The second example is a "paradox" cited in a Feynman's textbook on electromagnetism [35]. Imagine that we construct a device like that shown in Fig. 8-2.

There is a thin, circular plastic disk supported on a concentric shaft with excellent bearings, so that it is free to rotate. On the disk is a coil of wire in the form of a short solenoid concentric with the axis of rotation. This solenoid carries a steady current provided by a small battery, also mounted on the disk. Near the edge of the disk and spaced uniformly around its circumference are a number of small metal spheres insulated from each other and from the solenoid by the plastic material of the disk. Each of these small conducting spheres is charged with the same electrostatic charge q. Everything is quite stationary, and the disk is at rest. This is illustrated in Fig. 8-2.

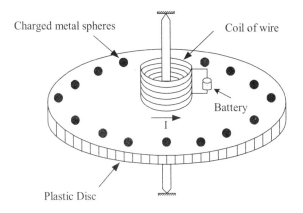

Feynman's paradox. Will the disc rotate if the current I is stopped?

Fig. 8-2

Suppose now that by some accident, or by pre-arrangement, the current is interrupted, without, however, any intervention from outside. So long as the current continues, there was a magnetic flux through the solenoid more or less parallel to the axis of the disk. When the current is interrupted, this flux must go to zero. There will, therefore, be an electric field induced, which will circulate around in circles, centered at the axis. The charged spheres on the perimeter of the disk will all experience an electric field tangential to the perimeter of the disk. The electric force is in the same sense for all the charges and will result in a net torque on the disk. From these arguments we would expect that as the current in the solenoid disappears, the disk would begin to rotate. If we knew the moment of inertia of the disk, the current in the solenoid and the charges on the small spheres, we could compute the resulting angular velocity.

But we could also make a different argument. Using the principle of the conservation of angular momentum, we could say that the angular momentum of the disk with all of its equipment is initially zero (aside from the negligible angular momentum of the current), and so the angular momentum of the assembly should remain zero. There should be no rotation when the current is stopped. Which argument is correct? Will the disk rotate or will not?

The solution is not easy, nor is it a trick. The answer in the standard electromagnetism involves the fields and their associated scalar and vector potential. There is an electromagnetic angular momentum in the surrounding empty space whose density is given by $\mathbf{r} \times \mathbf{A}$ where $\mathbf{A}$ is the vector potential due to the current in the solenoid. The correct answer is that the disk will rotate, since the electromagnetic angular momentum is transferred into the mechanical angular momentum of the system as the current stops [34, 36].

How does Weber's theory treat this problem? Consider that the charged metal spheres are located on a circle of radius R, coaxial with the solenoid that has a circular cross section of r and having n turns per unit length [36]. With respect to a rectangular coordinate system where the origin is at the center of the loop and the z axis is along the common symmetry axis, a typical sphere is located at $\mathbf{R} = R\mathbf{i}$ and an arbitrary point on a typical turn of the solenoid is located at $\mathbf{r} = r\cos\theta\mathbf{i} + r\sin\theta\mathbf{j} + z\mathbf{k}$. The charge on the sphere is Q and a charge on the solenoid turn at $\mathbf{r}$ is denoted by q. If n is sufficiently large, z will be nearly independent of θ. Let us consider the solenoid to consist of a stack of independent, parallel turns which carry the same current I. Given a force between Q and q, we will integrate over θ and z to find the force on the typical charged sphere.

Rewriting Weber's force of Eq. (7-12) for this situation,

$$\mathbf{F} = \frac{qQ}{4\pi\varepsilon_o s^2}\left(1 + \frac{s\ddot{s}}{c^2} - \frac{\dot{s}^2}{2c^2}\right)\hat{\mathbf{s}} \qquad (8\text{-}1)$$

where s is the separation of the charges in interaction. In the present case,

$$\mathbf{s} = (R - r\cos\theta)\mathbf{i} - r\cos\theta\mathbf{j} - z\mathbf{k} \qquad (8\text{-}2)$$

and

$$s = \left(R^2 + r^2 - 2Rr\cos\theta + z^2\right)^{1/2} \qquad (8\text{-}3)$$

and $\hat{\mathbf{s}} = \mathbf{s}/s$. We also have $\dot{s} = Rr\dot{\theta}\sin\theta/s$ and

$$s\ddot{s} = -\dot{s}^2 + Rr\ddot{\theta}\sin\theta + Rr\dot{\theta}^2\cos\theta .$$

As we need the component of the force tangential to the circle of radius R, we note that

$$\hat{s}\cdot j = -r\sin\theta/s \tag{8-4}$$

We can express q, the charge on a loop of the solenoid, as $n\lambda r d\theta dz$ in anticipation of the integrations over θ and z. For the stationary positive charges in the solenoid, Weber's force reduces to Coulomb's force and cancels the corresponding term in the force due to the moving electrons. The net tangential force on the charge Q is then given by

$$F = -\frac{Qn(-\lambda^-)r^3 R}{4\pi\varepsilon_o c^2}\int_0^L dz \cdot \left[\begin{array}{c}\dot{\theta}^2\int_0^{2\pi}\dfrac{\sin\theta\cos\theta d\theta}{(R^2+r^2-2Rr\cos\theta+z^2)^{3/2}}\\-\dfrac{3Rr\dot{\theta}^2}{2}\int_0^{2\pi}\dfrac{\sin^3\theta d\theta}{(R^2+r^2-2Rr\cos\theta+z^2)^{3/2}}\\+\ddot{\theta}\int_0^{2\pi}\dfrac{\sin^2\theta d\theta}{(R^2+r^2-2Rr\cos\theta+z^2)^{3/2}}\end{array}\right] \tag{8-5}$$

where n is the number of turns and λ^- is the linear electron density in each turn of the solenoid. The first two θ-integrals are readily shown to be zero, while the third can be evaluated in terms of the elliptic integrals:

$$F = -\frac{Qn(-\lambda^-)r^3 R\ddot{\theta}}{4\pi\varepsilon_o c^2}\int_0^{2\pi}\sin^2\theta d\theta\int_0^L\frac{dz}{(R^2+r^2-2Rr\cos\theta+z^2)^{3/2}} \tag{8-6}$$

Identifying $\dot{I} = -\lambda^- r\ddot{\theta}$ yields the force on q:

$$F = -\frac{Qnr^2 R\dot{I}}{4\pi\varepsilon_o c^2}\int_0^{2\pi}\frac{\sin^2\theta d\theta}{(R^2+r^2-2Rr\cos\theta+z^2)^{3/2}} \tag{8-7}$$

We find that the change in the current of the solenoid causes the force opposite to the change of the current.

The above examples show that both the conventional electromagnetic theory and Weber's theory are capable to explain the same electromagnetic phenomena equally well and prove that the great compatibility exists between the conventional electromagnetic theory with electric and magnetic fields and their associated potentials and Weber's electrodynamics with the relative coordinate between two charges and its derivatives.

Interaction of two moving charges

First, we consider two charged particles q_1 with mass m_1 and q_2 with mass m_2 that interact with each other through Weber's law. We can write the conserved energy in rest frame as

$$E = T + U = \frac{1}{2}m_1 v_1^2 + \frac{1}{2}m_2 v_2^2 + \frac{q_1 q_2}{4\pi\varepsilon_o r}\left(1 - \frac{\dot{r}^2}{2c^2}\right) \quad (8\text{-}8)$$

We analyze this problem in the inertial frame of reference in which the center of mass is at rest.

$$E = \frac{\mu(\dot{r}^2 + r^2\dot{\varphi}^2)}{2} + \frac{\alpha}{r}\left(1 - \frac{\dot{r}^2}{2c^2}\right) \quad (8\text{-}9)$$

where the reduced mass is given by $\mu = m_1 m_2/(m_1+m_2)$ and $\alpha \equiv q_1 q_2/4\pi\varepsilon_o$.

If the angular momentum $L = \mu r^2 \dot{\varphi} = 0$ (radial motion), we can easily solve this equation and obtain

$$\frac{\dot{r}}{c} = \pm\sqrt{\frac{2(rE - \alpha)}{r\mu c^2 - \alpha}} \quad (8\text{-}10)$$

If $E = \mu c^2$ the two charges will approach or move away from each other at a constant relative velocity $\dot{r} = \pm\sqrt{2}c$, for any r, as if they did not feel one another. We should remark here that to obtain this result we utilized the classical kinetic energy. The above result is only valid for radial velocities much less that the speed of light.

For $L \neq 0$, we can solve Eq. (8-2) for scattering problems. The result shows that scattering deflection angles are different for attractive and for repulsive central forces [37]. Note that the deflection angles in the classical Rutherford scattering are the same for attractive and repulsive central forces. In Eq. (6-11) of Chapter 6, we have seen that Ampere's force law modifies the Rutherford scattering angle. According to Weber's force, the difference in the deflection angles of attractive and repulsive forces was shown to be an increasing function of v^2/c^2. We are not aware of any experiment that tried to test these predictions.

Inertial mass of a charged particle in a potential

One of interesting situations is that of a charged spherical shell interacting with a point charge q. We follow the work by Assis [38]. Suppose a spherical shell of radius R, made of a dielectric (non-conducting) material, charged uniformly with a net charge q and spinning with an angular velocity $\omega(t)$ relative to an inertial frame S. The center of shell is located at the origin O of S. A point charge is located at the time t at **r**, and moves with velocity $\mathbf{v} = d\mathbf{r}/dt$ and acceleration $\mathbf{a} = d\mathbf{v}/dt = d^2\mathbf{r}/dt^2$ relative to the origin O as shown in Fig. 8-3.

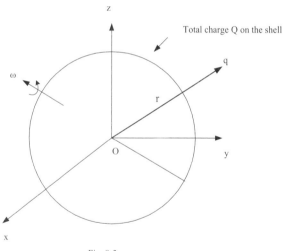

Fig. 8-3

We can integrate Weber's potential energy of Eq. (7-25) utilizing spherical coordinates. To do so, we employ $q_s = \sigma da = (Q/4\pi R^2)R^2\sin\theta d\theta d\phi$ where σ is the surface charge density. After the integration we obtain the potential energy for a moving charge q inside and outside of the shell as

$$U(r<R) = \frac{qQ}{4\pi\varepsilon_o}\frac{1}{R}\left[1-\frac{v^2-2\mathbf{v}\cdot(\boldsymbol{\omega}\times\mathbf{r})+(\boldsymbol{\omega}\times\mathbf{r})\cdot(\boldsymbol{\omega}\times\mathbf{r})}{6c^2}\right] \quad (8\text{-}11)$$

$$U(r>R) = \frac{qQ}{4\pi\varepsilon_0 r}\left\{1-\frac{1}{2c^2 r}[\mathbf{r}\cdot(\mathbf{v}-\boldsymbol{\omega}\times\mathbf{r})]\right.$$
$$\left. -\frac{1}{6c^2}\frac{R^2}{r^2}\left[(\mathbf{v}-\boldsymbol{\omega}\times\mathbf{r})\cdot(\mathbf{v}-\boldsymbol{\omega}\times\mathbf{r})-3[\hat{\mathbf{r}}\cdot(\mathbf{v}-\boldsymbol{\omega}\times\mathbf{r})]^2\right]\right\} \quad (8\text{-}12)$$

The associated Weber's force can be written as, for $r < R$,

$$\mathbf{F}(r<R) = \frac{\mu_o qQ}{12\pi R}\left[\mathbf{a}+\boldsymbol{\omega}\times(\boldsymbol{\omega}\times\mathbf{r})+2\mathbf{v}\times\boldsymbol{\omega}+\mathbf{r}\times\frac{d\boldsymbol{\omega}}{dt}\right] \quad (8\text{-}13)$$

We get a similar expression for $\mathbf{F}(r > R)$ that is omitted here.

An interesting result we can obtain is for a case where a point charge q moves inside the sphere and non-spinning charged spherical shell ($\boldsymbol{\omega} = 0$). If it is accelerated relative to the shell by other forces, Eq. (8-13) predicts that the shell will exert a force on q given by

$$\mathbf{F}_{\text{shell on }q} = \frac{\mu_o qQ}{12\pi R}\mathbf{a} \equiv m_W \mathbf{a} \quad (8\text{-}14)$$

where $m_W = \mu_0 qQ/12\pi R$ is what we call Weber's inertial mass for this situation.

Applying Newton's second law to this problem, we have

$$\sum_{j=1}^{N}\mathbf{F}_{jq} + \mathbf{F}_{shell-on-q} = m\mathbf{a} \quad (8\text{-}15)$$

In this equation $\mathbf{F}_{jq}$ is the force exerted by body j on q. The sum excludes the force exerted by the shell on q. Inserting Eq. (8-14) in Eq. (8-15) we have

$$\sum_{j=1}^{N}\mathbf{F}_{jq} = (m-m_W)\mathbf{a} \quad (8\text{-}16)$$

This shows that the test charge will behave according to Weber's electrodynamics as if it had an effective inertial mass given by $m - m_w$. If q and q are of the same sign, then $m_w > 0$. In principle we might increase m_w by increasing q/R, so that it might become eventually equal to m. The velocity of the effective mass would become infinite for any force! This was once argued by Helmholtz in 1872 to refute Weber's law [39]. Helmholtz argued against Weber's theory that would violate the principle of the conservation of energy. His criticism can be answered by at least two considerations. First, the inertial mass derived above is based on Newtonian mechanics and the cases of high velocity are not valid. As we discussed after Eq. (8-10), Weber's theory shows that the motion of a particle interacting with another has the limit for its velocity $v < \sqrt{2}c$, where c is the speed of light. Second, by use of both Weber's electrodynamics and Schrodinger's gravitational interaction, it is shown that the particle's velocity never exceeds c under Helmholtz's condition [40].

Is this a failure of Weber's electrodynamics? Only experiments will determine the prediction by Weber's law. How can we test this? One of possible situations of testing is conceived as follows.

Let us suppose the test charge is an electron ($m = 9.1 \times 10^{-31}$ kg, $q = -e$ $= -1.6 \times 10^{-19}$ C). Choosing the zero of the potential of the shell at infinity, Weber's inertial mass can be written as $m_w = qV/3c^2$, where $V = q/4\pi\varepsilon_o R$ is the potential of the shell. In order to make $m_w = m$, the electron would need to be inside a spherical shell charged to 1.5×10^6 volts. It is not impossible to obtain potentials of this order of magnitude. The difficulty of such an experiment would be discharges to surroundings. The required high voltage causes dielectric breakdown in air, and any conducting lead for a device to accelerate electrons in the shell would become a passage for discharge. For experimental evidence, though it is not yet verified, Mikhailov published his work that showed the effective mass changed according to Eq. (8-16) [41]. This work may be a landmark, if his results are confirmed by other workers.

Weberian induction

Another test we propose is a variation of the above experiment. This test involves a new kind of induction that is predicted by Weber's electrodynamics but not by Maxwellian electrodynamics [42]. The new induction happens when a metallic disc rotates at a constant angular speed about its axis of symmetry in a uniformly charged shell. Suppose the disk has a radius r and rotates at a constant angular speed ω relative to the laboratory by external mechanical means. This disk is placed in a charged spherical shell with radius R as discussed in the previous example. The potential energy of a charged particle q inside the charged shell is given by from Eq. (8-11) as

$$U = \frac{qQ}{4\pi\varepsilon_o R}\left(1 - \frac{v^2}{6c^2}\right) \qquad (8\text{-}17)$$

Note that according to Lorentz's force and Maxwell's equations, the corresponding potential is, for outside,

$$U = \frac{qQ}{4\pi\varepsilon_o R} \qquad (8\text{-}18)$$

and the force on q inside is zero, different from the Weber's force on q given by Eq. (8-14). As the free electrons in the disk have a negative charge and the centripetal acceleration points radially towards the origin, this Weberian force will point radially outwards if $q > 0$ (or radially inwards if $q < 0$). This means that according to this force the disk will tend to become negatively charged at he center if $q > 0$. In the steady state situation there will be created a Weberian electric field $\mathbf{E}_w$ which will balance this Weberian force, that is, $q\mathbf{E}_w = -\mu_o q Q \mathbf{a}/12\pi R$. This Weberian electric field is due to the polarized charges generated by Weber's force.

The potential difference (voltage) between the periphery and the center of the disk is given by

$$\Delta\phi = \phi(r) - \phi(0) = -\int_{\pi=0}^{r} \mathbf{E}_w \cdot d\rho = -\frac{\mu_o Q \omega^2 r^2}{24\pi R} \qquad (8\text{-}19)$$

where we used the centripetal acceleration $a = -\omega^2 \rho$.

Choosing the zero potential at infinity, the potential of the charged sphere can be written as $\phi_s = Q/(4\pi\varepsilon_o R)$. Then, the Weberian induction can be written as

$$\Delta\phi_W = -\phi_s \frac{\omega^2 r^2}{6c^2} \qquad (8\text{-}20)$$

APPLICATIONS OF WEBER'S ELECTRODYNAMICS

If $\omega = 100\pi$ radians/s and $r = 10$ cm, we obtain $\Delta\phi_w = -1.8 \times 10^{-15}\ \phi_s$. This is very small voltage for low voltage of the sphere. Again, if we try the maximum voltage of the sphere to be 1.5 MV, the potential difference by Weberian induction would be 2.7×10^{-9} volts. An experimental verification will require a very difficult task to detect the minute effect among other larger effects including induction by the earth's magnetic field.

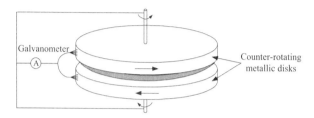

Fig. 8-4. Detection of Weberian induction

For detection of the new effects that are only predicted by Weber's electrodynamics, we propose the following experimental set-up illustrated in Fig. 8-4.

We should note that there occur additional voltages induced by other mechanisms. One is inertial induction that has the same magnitude of the Weberian induction. The inertial induction that occurs regardless of the existence of an external uniform potential is not bad since it can be distinguished from the Weberian induction that occurs inside the charged sphere. The other induction is due to the earth's magnetic field. The magnitude of the induction due to the earth's magnetic field is estimated to be 78 mV for $B_E = 5 \times 10^{-5}$ T, far larger than the Weberian's or inertial inductions. This magnetic induction, however, changes sign according to the direction of rotation, while the Weberian and inertial inductions do not change their signs. We can cancel the magnetic induction by utilizing a pair of counter-rotating disks. If we connect the two disks at their central parts and their rims by a metal brush, the magnetic effect should disappear. We can detect the minute effect due to the Weberian induction that depends on ω^2. The validity of Weber's electrodynamics can be tested by an experiment to check this new effect.

Chapter 9. The propagation of electrical signals in vacuum

We have shown in Chapter 7 that Weber's theory is compatible with Maxwell's theory as to the basic laws of Ampere's and Faraday's. The main shortcoming of Weber's theory, however, is that the theory cannot apparently describe electromagnetic radiation. This incompleteness is often considered "fatal" as a theory to describe the electromagnetic phenomena. The Weber force, however, can lead to the telegrapher's equation for a conducting medium, and the propagation of the electrical signal is shown to travel at the speed of light [Kirchhoff 1857]. Assis re-derived the same equation in modern terms [43]. In his derivation, charged particles of an electrical signal interact with the charges readily available in the conductor. Using Weber's force and Newtonian mechanics, the signal is found to travel at the speed of light, resembling a similar mechanism of sound wave propagation in a material medium.

Our concern is if the vacuum could provide the necessary medium for application of Weber's theory. The concept of vacuum has changed considerably by the development of quantum mechanics. We shall review the idea of the vacuum in the following [44].

Review of vacuum

Electromagnetic waves and light travel in free space at the speed of light c. The free space is the vacuum. What is, then, the vacuum? Greek philosopher Democritus in the fifth century considered the vacuum as the void. The void was necessary to explain his idea about the motion of atoms. Aristotle opposed the idea of invisible atoms and a total void from his philosophical viewpoint. Aristotle filled the vacuum with ether. The ether was the stuff of stars and the heavens, but it also permeated the four elements — earth, fire, air, and water — of the lowly world. Even when the four elements turned out not to be elemental, the concept of ether survived for more than 1000 years since Aristotle. It became an essential component of nineteenth century physics, when light was found to consist of electromagnetic waves. Waves need a medium to travel. Physicists reasoned space could not be empty. It must be filled with an ether. The ether, though, was strange stuff. Since the speed of light is so tremendously high – 186,000 miles per second, the ether had to be exceedingly firm, yet bodies move through it without encountering detectable resistance. The physicists of a century ago could see no way to do without it.

Around the turn of the century, however, the ether's foundations were shaken. In 1887, Michelson and Morley performed an experiment to prove the existence of the ether. They found nothing. In 1905 Albert Einstein declared simply that the ether hypothesis was superfluous and his famous special theory of relativity was born. The vacuum becomes empty again. But then, a quarter of century later it began to fill up again — this time with the conceptual fruits of quantum mechanics. Quantum mechanics was introduced in 1925 as a replacement for Newton's mechanics in the description of atomic phenomena such as the emission and absorption of light. It turned out to have implications for the vacuum as well. The key concept here is the uncertainty principle that prohibits the position and the speed of a particle to be determined with certainty at the same time. One consequence is that particles and other system in motion have what is called a zero point energy even at the absolute zero temperature. Zero point energy affects the vacuum state. An oscillating system of charged particles emits electromagnetic waves.

A vacuum is then inevitably filled with electromagnetic fluctuations of all wavelengths. quantum mechanics predicts a phenomenon even more exotic than electromagnetic vacuum fluctuations. Occasionally a fluctuation carries enough energy to materialize into a pair of new particles. All of a sudden, for a brief moment, a negative electron and its antimatter twin, a positron, pop up out of nowhere. Together they preserve the electrical neutrality of the vacuum, and in an instant they annihilate each other and vanish without trace. If there happens to be a strong electric field, however, the electron and the positron will respond, so during its brief lifetime the pair may line up to the field. Thus the vacuum becomes momentarily polarized. Note that the positron, the antiparticle of the electron has a positive mass equal to that of the electron. When a pair of a particle and its antiparticle pops up, their created mass energies can be canceled by the attractive potential energy that is negative for opposite electrical charges of the pair. Any discrepancy in energy is, however, possible as long as it satisfies the range allowed by the uncertainty principle.

Electric signal propagation in vacuum

Now a modern view of vacuum is filled with pairs of all kinds of elementary particles and their anti-particles, positives and negatives, keeping it electrically neutral. If an electrical disturbance occurs at a point of space that is readily polarizable, we can imagine a charge separation formed immediately in the vicinity of the source. If we assume, for simplicity, that the disturbance creates a charge separation in the shape of a coaxial cable for an infinitesimal distance of say, dx. We may call this a "virtual' coaxial cable of length dx. If the disturbance is circularly polarized electromagnetic wave, the situation may indeed be like this way. The vacuum may be considered as an ideal "dielectric" substance and yet without resistance, since it is filled with pairs of positive and negative particles with negligible masses. Thus, we consider electric signals propagating in the vacuum medium that has an equivalent circuit consisting of a series of LC components (per dx) connected to infinity shown in Fig. 9-1.

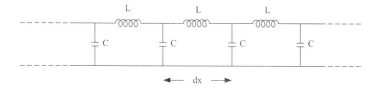

Fig. 9-1. LC circuit for the telegraph equation

The result is the telegrapher's equation that describes the signal propagation with a speed $v = 1/\sqrt{LC}$, where we can calculate L and C for a coaxial configuration. We obtain, from either Weber's theory or Maxwell's theory, $L = (\mu_o/2\pi)\ln(b/a)$ and $C = 2\pi\varepsilon_o/\ln(b/a)$ where constants a and b are arbitrary. The propagation speed is readily seen as $v = 1/\sqrt{\varepsilon_o \mu_o} = c$, the speed of light in free space. This is analogous to the propagation of sound waves, as we can derive the sound propagation from Newton's laws. Therefore, it is possible to add the propagation of electrical signals in space according to Weber's theory by implementing some aspects of modern view of the vacuum. We should note here that Weber's theory requires two charged particles to interact and if no receiving particle exists, signals will not propagate — no electromagnetic waves will occur. This is an important conceptual difference between Weber's theory and Maxwell's theory for the existence of electromagnetic waves.

Bending of light near a massive body

It is interesting to extend our new view of the vacuum to a possible alteration of free space in the vicinity of a massive body. We will discuss such a possibility.

In vacuum, pairs of particle and antiparticle pop out at any moment and everywhere. Since antiparticles have positive masses as regular particles do, the instantaneous density of pair existence could be affected by the presence of a massive body due to its gravitational attraction. It would be conceivable to allow the pair density to vary as a function of radius. The density of the pairs' electric dipoles would become higher in space closer to the body. Then, the dielectric property or the permittivity of space could vary with radius. What would this imply for the propagation of light? The speed of light would vary as a function of radius! Moreover, we could expect that light would bend around a massive body just like the general theory of relativity predicts. How, then, can we argue for determining the permittivity in free space as a function of distance from a massive body? There is a good work by Li and Tian on the light speed in a gravitational field [45]. Their calculation is based on the conservation of energy and momentum. They showed that the speed of light changes as

$$c = c_o \left(1 - \frac{2GM}{rc_o^2}\right)^{1/2} \tag{9-1}$$

where c_0 is the speed of light in free space, M is the mass of a nearby star and G is the gravitational constant. The index of refraction is then $n = c/c_0$.

We will describe here a unique approach to find the index of refraction. Without a specific prescription of the vacuum, Hayden derived a result for the dependency of the permittivity on the distance from a massive body. He applied the law of the conservation of energy to a pair of positive charges separated at a distance near a massive body, and used it to predict the deflection of starlight [46].

According to his method, we consider two positive charges with the same charge q, separated at a distance D, and at some distance R from the center of, *e.g.*, the sun. For this pair of charges we could perform a perpetual cycle that has four processes; 1) apply an external force to bring the charges close together at one radius, 2) lifting them to a larger radius, keeping the distance constant, 3) allowing the forces to move apart to the original separation, and 4) dropping the charges to the original radius. This is illustrated in Fig. 9-3.

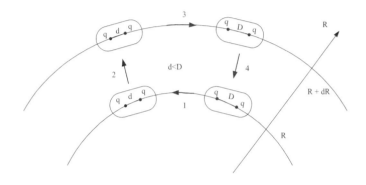

Fig. 9-2

Step 3 returns more energy than is done in step 1 because the permittivity is higher at lower radius. We will include the gravitational attraction of the electrostatic energy's equivalent mass to the body. We adopt the convention that the permittivity of free space ε_o is multiplied by the dimensionless relative permittivity ε_r, the latter being equivalent to a dielectric constant for a material medium. The relative permittivity ε_r is a function R.

We calculate the work done by an external force in the four steps beginning with that required to reduce the distance between the charges to a shorter distance d (from D of the original distance).

$$W_1 = \frac{q^2}{4\pi\varepsilon_o} \frac{1}{\varepsilon_r} \left(\frac{1}{d} - \frac{1}{D} \right) \tag{9-2}$$

In the second step the assembly is raised an amount dR. The work done is

$$W_2 = GM \left(m + \frac{q^2}{4\pi\varepsilon_o \varepsilon_r d} \frac{1}{c^2} \right) \frac{dR}{R^2} \tag{9-3}$$

where we assume that a dimensionless factor ε_r varies slowly with R. The mass M is the mass of the sun. The mass m is any mass associated with the assembly of the charge pair and will soon cancel out. The term added to m is the equivalent mass of the electrostatic energy. In step 3 charges separate to their original separation D, the work is

$$W_3 = \frac{q^2}{4\pi\varepsilon_o} \frac{1}{\varepsilon_r{'}} \left(\frac{1}{D} - \frac{1}{d} \right) \tag{9-4}$$

Here the permittivity takes the value $\varepsilon_r{'}$. As we expect $\varepsilon_r{'}$ to be less that ε_r, W_3 is larger than W_1 in magnitude. Finally step 4 puts the system back to its original configuration.

$$W_4 = -GM \left(m + \frac{q^2}{4\pi\varepsilon_o \varepsilon_r{'} D} \frac{1}{c^2} \right) \frac{dR}{R^2} \tag{9-5}$$

We now impose two constraints: First, the total work done for the loop is zero,

$$W_1 + W_2 + W_3 + W_4 = 0 \tag{9-6}$$

and second that the magnetic permeability $\mu_0\mu_r$ has the same dependence upon R and M as the permittivity. The latter constraint assures that the impedance, $(\eta/\varepsilon)^{1/2}$, remains constant, and that dispersion cannot occur. We obtain after some algebra, with $d\varepsilon_r = \varepsilon_r{'} - \varepsilon_r$,

PROPAGATION OF SIGNALS

$$\frac{d\varepsilon_r}{\varepsilon_r} = -\frac{GM}{c^2}\frac{dR}{R^2} \qquad (9\text{-}7)$$

Constraint 2 requires that c be c_o/ε_r, so our differential equation becomes

$$\frac{d\varepsilon_r}{\varepsilon_r^3} = -\frac{GM}{c_o^2}\frac{dR}{R^2} \qquad (9\text{-}8)$$

The result is

$$\varepsilon_r = \left(1 - \frac{2GM}{c^2 R}\right)^{-1/2} \qquad (9\text{-}9)$$

Since n (index of refraction) $= (\varepsilon_r \mu_r)^{1/2}$, thus, $\varepsilon_r = c_o/c$ in this case since μ_r has the same dependency. We have for the speed

$$c = c_o \sqrt{1 - \frac{2GM}{R c_o^2}} \qquad (9\text{-}10)$$

Tangherlini also obtained the same result for the index of refraction by considering a photon as classical particle by classical mechanics [47].

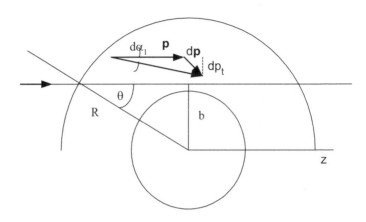

Fig. 9-3. Deflection of light momentum by gravity

Bending of light due to a mass as it passes by is conventionally reasoned by the general theory of relativity. Equivalently, we can calculate the deflection by the "classical" reasons as follows. The idea is that light with energy E has the equivalent mass E/c^2, and its momentum E/c is deflected by gravity. The deflection is illustrated in Fig. 9-4.

Light with impact parameter b has momentum E/c when at R, but receives an impulse dp along the radius in moving from R to $R - dR$. The component of the impulse perpendicular to the velocity is obtained by multiplying by $\sin\theta$. The deflection angle $d\alpha_1$ is the transverse momentum divided by the momentum, dp_t/p, where the former is

$$dp = Fdt = -\frac{1}{c}\frac{GM(E/c^2)}{R^2}dR \qquad (9\text{-}11)$$

We obtain
$$d\alpha_1 = \frac{GM}{c^2}\frac{bdz}{\left(b^2+z^2\right)^{3/2}} \qquad (9\text{-}12)$$

which is shortly to be integrated. b is the impact parameter, and in this case, it is the distance of closest approach.

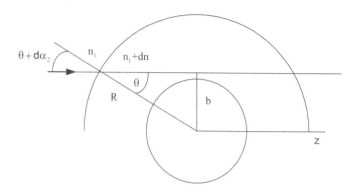

Fig. 9-4. Deflection of light by gravity

The new method of calculating the deflection is to use the dependency of light speed as function of R that was derived here. Note that the index of refraction is $n = \varepsilon_r$. The deflection is illustrated in Fig. 9-5. We apply Snell's law at the distance at R.

$$n_1 \sin(\theta + d\alpha_2) = (n_1 + dn) \sin \theta. \qquad (9\text{-}13)$$

Expanding, and using $\cos(d\alpha_2) \sim 1$, and $dn = (dn/dR)(dR/dz)dz$, we obtain

$$d\alpha_2 = \frac{\tan \theta}{n} \frac{dn}{dR} \frac{dR}{dz} dz \qquad (9\text{-}14)$$

from which we obtain after calculating Eq. (9-10) to first order and differentiating

$$d\alpha_2 = \frac{GM}{c^2} \frac{b\,dz}{\left(b^2 + z^2\right)^{3/2}} \qquad (9\text{-}15)$$

which is the same as Eq. (9-12), though for different reasons. We obtain the total deflection angle by summing Eqs. (9-12) and (9-15) and integrating from $z = -\infty$ to 0, and doubling. Letting $b = z \tan \theta$, and letting θ run from 0 to $\pi/2$, we obtain

$$\alpha_{total} = \frac{4GM}{bc^2} \tag{9-16}$$

where c is the regular constant speed of light. The same result is also obtained by Li and Tian [45]. This result of Eq. (9-16) is in agreement with experiment [48]. General relativity theory [49, 50] obtains the same result at the expense of more mathematics.

The deflection of starlight by the sun was measured in 1919 as a check on general relativity theory (GRT). The agreement thus obtained was heralded as support for GRT; "curved space-time" has since become a part of the literature. The problem solved by GRT was, however, one created by the special theory of relativity (SRT) whose principle is based on the definition that the speed of light is a universal constant.

We showed here that the deflection of light could be calculated using Newtonian time and Euclidean geometry without the "curved space-time" formalism. Moreover, the dependence of the index of refraction due to gravity expressed by Eq. (9-10) might give a clue to partially resolve the *dark matter* mystery. While carefully measuring the speed of rotation of galaxies, astronomers could estimate what the rotation should be by calculating the mass of all the visible stars and gas. Much to their surprise, the measurement showed that most galaxies are rotating much faster than they should. This meant that, according to Einstein's theory of relativity, these galaxies should be flying apart. Yet clearly, they are not. Einstein's theory of gravity cannot be wrong and scientists speculate that most galaxies are surrounded by some unknown dark matter that cannot be observed. Candidates for dark matter include MACHOS (MAssive Compact Halo Objects), WIMPS (Weakly Interacting Massive Particles), GAS (hydrogen gas) and others. Bad news is that there is no evidence for any of the above.

Epilogue

Concluding remarks

Starting from an intuitive and naïve idea about the interaction of two electric charges in motion, we have explored and obtained new formulas for their forces. The "new" formulas, however, turned out to be an old one. We encountered now forgotten theories that appeared in the middle of nineteenth century. Notably, Weber's theory is an interesting one that takes into consideration of only the relative coordinates of two interacting particles. Fundamentally, it is a relativity theory with an open end. Through studying those forgotten theories we could elucidate the position of the main classical theory in relative perspectives.

Classical electromagnetic phenomena are systematically explained by electromagnetic theory based upon Maxwell's equations together with Einstein's special theory of relativity. The abstract concept of the electric and magnetic fields play a major role in the theory. Charged particles can "directly" interact with each other through the go-between fields extended in space. The velocities of particles are defined by an inertial frame of an observer's choice. So, the interaction of two charged particles depends upon the choice of a reference frame, or an observer.

On the other hand, Weber's theory explains electromagnetic phenomena from the viewpoint based upon the relative motion of interacting particles. No fields are necessary. Forces between charged particles in relative motion play the major role in the theory. The electromagnetic phenomena can be explained as the result of the superposition of all the forces between charged particles. This approach may be intuitively more appreciated than that of using abstract fields.

We have shown that Weber's theory can explain electromagnetism as well as Maxwell's theory. Weber's theory, moreover, can predict some phenomena that are not predicted by Maxwell's theory. The predictions are to be verified by an experiment. The experiments to test the predictions are very difficult since the predicted effects are either extremely small or the setups for experimenting are very costly. So far, the predictions are not yet verified. Weber's theory, thus, presents no better than the conventional theory. The conventional theory of Maxwell's equations with the special theory of relativity has been the paradigm of classical electromagnetism.

Looking back, the classical physics in the nineteenth century — physics based on Newtonian mechanics, thermodynamics, and electricity and magnetism — explains the natural phenomena and physical behaviors that can easily be observed. Classical physics agrees with our intuition, borne of experience. In the latter part of the nineteenth century, many reputable scientists believed that knowledge of all important areas of physics had attained virtual completion — the end of physics (A. A. Michelson in 1894; Lord Kelvin in 1900) [51].

Before this optimistic view prevailed, the battle to attain superiority between Maxwell's theory and Weber's theory was carried out among European physicists. A decisive blow to Weber's theory was also a triumphant discovery that Maxwell's theory based on the concept of field could predict the existence of electromagnetic wave in free space. Relativity theory and quantum mechanics not only nicely encompassed the field theory, but also leaped to establish a new paradigm of physics. Weber's theory attempted to reduce all events in nature to forces acting between material particles, including electrically charged particles. Before the discoveries of elementary particles such as the electron and the proton, Weber's view was astonishingly modern. By Weber's theory, we have shown easier and intuitively make-sense solutions for many difficult electromagnetic problems than by the standard theory. We examined the possible effects that Weber's theory predicts but Maxwell's theory does not. The unsuccessful prediction of the electromagnetic wave propagation in free space, once considered a fatal defect for Weber's theory, can be mended. In chapter 9, we explored a possible explanation of the propagation of electromagnetic waves in vacuum by Weber's theory by adapting the modern view of the vacuum. Furthermore this modern view could lead to a simpler and easier understanding of the light bending by a massive body. Only further experiments can test the validity of Weber's theory.

Since 1872 when Helmholtz declared that Weber's theory was heretical, refereed physics journals have shied away from publishing papers that are based on Weber's theory. We have shown that the objections raised by Helmholtz are easily answered. His criticism was a confusion caused, perhaps, by mixing non-relativistic kinetic energy and a basically relativistic potential energy [39]. It seems that Weber's theory is heretical in the sense that theory does not conform to the present paradigm of electromagnetic theory, established based on Maxwell's theory and the special theory of relativity. In contrast to the general dismissal of Weber's theory as heretical, we tried to show that Weber's theory is much better than people had regarded.

We have witnessed the incompleteness of Weber's theory, in which the iterations to higher derivatives of relative position are not performed. On the other hand, Maxwell's theory together with the Einstein's special theory of relativity is so self-consistent and complete within, and that has become the present paradigm. Despite its success, the special theory of relativity has continued to be a target of criticism.

Further implications

Weber's theory also has a significant implication for Mach's principle applied to gravitation. This aspect of Weber's theory is well discussed by Assis's book on "Relational Mechanics" [52].

According to Newtonian mechanics there are motions of bodies relative to free space and we can detect these motions when the bodies are accelerated (relative to absolute space, as Newton put it, or relative to inertial frames of reference, as we would say today). For instance, how can we know that the earth really rotates around the north-south axis with a period of one day? Foucault's pendulum gives evidence of this effect. Let us consider the simplest case of a pendulum swinging at the North Pole of the earth. The plane of oscillation does not remain fixed relative to the earth but precesses with a period of one day. In classical mechanics this is interpreted as due to the earth's absolute rotation, without any relation with the distant universe (stars and galaxies).

Leibniz, Berkeley and Mach rejected the concept of absolute space and proposed that there are only motions of bodies relative to other bodies. Accordingly only these relative motions between material bodies could be detected or lead to measurable effects. This idea became known as Mach's principle. From this point of view the precession of Foucault's pendulum is due to the rotation of the earth relative to the distant material universe. If we could keep the earth stationary and rotated the distant universe around the north-south axis of the earth in the opposite direction with a period of one day, a pendulum swinging at the North Pole should precess following the rotation of the universe. This means that we cannot distinguish between the two situations, as both of them lead to the same observed effects. This is the essence of Mach's principle [53]. And if there was no relative rotation between the earth and the distant universe, the pendulum should not precess. That is, if the rotation of the distant universe relative to the earth could be stopped, the precession of the pendulum should also stop. Since we cannot control the motion of the distant universe, our idea here is to explore the consequences of Mach's ideas in electromagnetism.

There is an effect analogous to Foucault's experiment when we deal with classical electromagnetism. We will perform a thought experiment with a charged pendulum. Let us suppose that there is a pendulum of mass m and length ℓ oscillating in a vertical plane due to a uniform gravitational field g. The frequency of oscillation for small amplitudes is given by $\omega = \sqrt{g/\ell}$. If we are in an inertial frame of reference, the plane of oscillation of the pendulum will not precess. Now consider a charge q attached to the mass of the pendulum and place it in an uniform magnetic field $\mathbf{B}$ pointing upwards. The plane of oscillation will precess relative to the inertial frame of reference with an angular velocity given by $\Omega = -qB/2m$, supposing a weak magnetic field such that $|qB/m\omega| \ll 1$ (Ref. 49, p. 45). The negative value of Ω indicates a rotation in the clockwise direction when the pendulum with a positive oscillating charge is seen from above.

There are three basic ways of creating a uniform magnetic field: the region near the poles of a large magnet, the region inside a long coil carrying a constant current (or equivalently the region near the center of Helmholtz's coils), or the region inside an uniformly charged spherical shell spinning with a constant angular velocity. In order to make the analogy with Foucault's experiment we will consider the magnetic field due to the spinning of a charged spherical shell. Let us suppose that a spherical shell of radius R with uniformly distributed charge q spins with a constant angular velocity $\boldsymbol{\omega}_Q$ relative to an inertial frame of reference. According to classical electromagnetism [54], this system creates a dipole magnetic field outside the shell and a constant and uniform magnetic field anywhere inside the shell given by $\mathbf{B} = \mu_o Q \boldsymbol{\omega}_Q / 6\pi R$, where $\mu_o = 4\pi \times 10^{-7}$ H/m is the magnetic permeability of the vacuum. The precession of the plane of oscillation of the charged pendulum inside this shell relative to the inertial frame of reference will have the angular velocity $\Omega = -\mu_o qQ\omega_Q / 12\pi mR$. When $qQ > 0$

($qQ < 0$) then $\Omega\omega_Q < 0$ ($\Omega\omega_Q > 0$), indicating rotations in opposite (the same) directions. From this expression we can see that if $\omega_Q = 0$, then $\Omega = 0$. This would be analogous to stopping the rotation of the distant universe in Foucault's experiment.

For a more experimentally plausible case let us replace the pendulum by a conducting ring of radius r placed at rest inside the spinning charged shell above, so that the axis of the ring is parallel to the rotation axis and to the uniform magnetic field **B** created by the shell. If this magnetic field is constant in time, no induction will happen in the ring. But if the magnetic field changes in time, there will be an electromotive force, *emf*, arising in the ring according to Faraday's law. It is given by $emf = -d\Phi/dt$, where $\Phi = \iint \mathbf{B} \cdot d\mathbf{a} = B\pi r^2$ is the flux of the uniform magnetic field across the area of the ring, with $d\mathbf{a}$ being an element of area. This electromotive force can be detected by the induced current generated in the ring. A change in the magnetic field can be accomplished by changing the rate of rotation of the shell, accelerating or decelerating it. The electromotive force in this case will then be given by $emf = -\pi r^2 dB/dt = -(\mu_o r^2 Q(d\omega_Q/dt))/6R$.

According to Mach's principle the same effect (*emf* and induced current in the ring) should happen if, instead of changing the rate of rotation of the shell, we change the rate of rotation of the ring with the opposite value. That is, if the charged spherical shell remains with a constant angular velocity and there is a change in the angular rotation of the internal ring given by $d\omega_r/dt$, there will be on it an induced *emf* given by: $emf = (\mu_o r^2 Q(d\omega_r/dt))/6R$. This effect is independent of the value of the angular rotation of the external shell, including the case of a stationary shell. The conventional electromagnetic theory does not predict this effect. In principle it can be tested in the laboratory.

Another way to change the magnetic flux without changing the rate of rotation of the spinning shell is to change the charge in the shell. For instance, by discharging the spinning shell the magnetic flux through the ring rapidly decreases. An *emf* is induced in the ring according to Faraday's law.

Mach's principle implies that only the relative motion between the shell and the ring matters. Therefore, when the charged sphere remains stationary and the ring rotates with a constant angular velocity, the same induction should occur in the ring when the shell is discharged. Induction occurs without apparent magnetic field. This effect has not yet been observed. Once more, it can presumably be tested in the laboratory.

We have elucidated the crucial discrepancies between the conventional electromagnetic theory and electromagnetic situations implied by Mach's principle. The new effects described above test the validity of the implications of Mach's principle to electromagnetism. Only experiments can decide how nature works.

In the present paradigm of science, the special theory of relativity and the general theory of relativity are the "correct" theories for the physics of the world. The speed of light is "defined" to be a universal constant. Until experiments prove otherwise, it is believed to be true. Lorentz's force with Lorentz's transformations is a "complete" description of the force on a charged particle. The general theory of relativity is the theory of gravitation and the curved spacetime is the dynamical arena of gravity. The path of light is along the curved path of spacetime. Predictions based on the general relativity are well documented. It is again said that no experimental evidence proves the theory wrong. The special relativity and general relativity are so complete within and constitute the present paradigm. Some people even argue that nature behaves according to the paradigm. That is arrogant and even dangerous.

Finally, I would like to present the following statement that is endorsed by the Council of the American Physical Society.

Science is the systematic enterprise of gathering knowledge about the world and organizing and condensing that knowledge into testable laws and theories.

The success and credibility of science is anchored in the willingness of scientist to:

(1) expose their ideas and results to independent testing and replication by other scientists; this requires the complete and open exchange of data, procedures and materials;

(2) abandon or modify accepted conclusions when confronted with more complete or reliable experimental evidence.

Adherence to these principles provides a mechanism for self-correction that is the foundation of the credibility of science.

APPENDIX A:
Notable Physicists on Electricity and Magnetism in the 19th Century

Physicist/year	1700	1750	1800	1850	1900
American			Henry |--------------------------|	
British			Faraday |-------------------------|	
				Maxwell |-------------------|	
				Heaviside |--------------------|
Danish			Oersted |-------------------------|	
French		Coulomb |-------------------------|		
			Ampere |-------------------|	
German			Gauss |-------------------------|	
			Weber |-------------------------------|	
			Helmholtz	|--------------------------|
				Kirchhoff |---------------------|	
				Hertz |---------------|
(American)					Einstein |-----------------
Dutch					Lorentz |----------------------|
Italian		Volta |-------------------------|		
year	1700	1750	1800	1850	1900

APPENDIX B: Biographies of notable physicists in the 19th century

Ampere, Andre-Marie (1775-1836), was a French physicist, mathematician, chemist and philosopher who is famous for founding the science of electromagnetics (which he named electrodynamics) and who also gave his name to the unit of electric current.

Ampere was born in Polemieux, near Lyons in France in 1775. The son of a wealthy merchant, he, instead of going to schools, was tutored privately and was, to a great extent, self-taught. His genius was evident from an early age, particularly in mathematics, which he taught himself and mastered to an extremely high level by the age of 12. The later part of his youth, however, was severely disrupted by the French Revolution. In 1793 Lyons was captured by the Republican army and his father — who was both wealthy and a city official — was guillotined. In 1802 he wrote his first important work on the mathematical theory of games and he was appointed Professor of Physics and Chemistry at the Ecole Centrale in Bourg. In 1809 he was appointed Professor of Mathematics. His talent had been recognized by Napoleon, who in 1808 he appointed him Inspector-General of he newly formed university system, a post he retained until his death in 1836.

Ampere studied a wide range of subjects, including psychology, philosophy, physics, and chemistry. Despite considerable and varied achievements, Ampere's fame today rests almost entirely on his work on electromagnetism, a discipline that he was responsible for establishing. His work in this field was stimulated by the finding of the Danish physicist Hans Christian Oersted that an electric current can deflect a compass needle. When Ampere witnessed a demonstration of this experiment, he, like other scientists, was prompted to hectic activity. Biot and Savart interpreted Oersted experiment as showing that the electric current had magnetized the wire which then interacted with the magnetic needle. Ampere, on the other hand, looked at the experiment differently. He thought that what was basic in the phenomenon was the direct interaction between currents, which meant that there should exist microscopic currents within the magnets. He showed that two parallel wires carrying current in the same direction attract each other, whereas when the currents are in opposite directions, mutual repulsion results. After several years of complete involvement of the experiments, in 1823 he arrived at the following force, dF, between the current elements Ids and $I'ds'$

$$dF^A = -\frac{II'dsds'}{r^2}\left(\frac{1}{2}\frac{dr}{ds}\frac{dr}{ds'} - r\frac{d^2r}{dsds'}\right)$$

This is called as Ampere's force law, and where I and I' are in the electrodynamic system of units. The force is along the line joining the elements of length ds and ds', and r is the distance between them. He established precise mathematical formulations of electromagnetism, a relationship of the magnetic force arising between two current-carrying conductors with their currents and the distance between them. This relationship is known as Ampere's law — in today's term, an equation that relates the magnetic field and the current that produces it, or

$$\oint \mathbf{B} \cdot d\ell = \mu_0 I_{enclosed} \qquad \text{Ampere's law}$$

Ampere's work gave a great impact and stimulated much further research into electromagnetism.

Coulomb, Charles (1736-1806), was a French physicist who established the laws governing electricity and magnetism. The unit of electric charge is named the coulomb in his honor.

Coulomb was born in Angouleme in France in 1736 and educated at the Ecole du Genie in Mezieres, graduating in 1761 as a military engineer with the rank of First Lieutenant. In 1774, Coulomb had become a correspondent to the Paris Academy of Science. In Paris, his duties were those of an engineering consultant and he had time in hand for his physics research. The year 1789 marked the beginning of the French revolution and Coulomb found it prudent to resign from the army in 1791.

Coulomb's major contribution to science was in the field of electrostatics and magnetism, in which he made use of a torsion balance he invented in 1777. He was able to show the torsion suspension can be use to measure extremely small forces. Coulomb was interested in the work on electrical repulsion and, in 1785, using his torsion balance he demonstrated the force between two bodies opposite charge is directly proportional to the product of the charges on each, and inversely proportional to the distance between them. The results are embodied in Coulomb's law, in the modern terms,

$$F = \frac{1}{4\pi\varepsilon_o} \frac{q_1 q_2}{r^2} \qquad \text{Coulomb's law}$$

where ε_o is the vacuum permittivity.

With his researches on electricity and magnetism, Coulomb brought this area of physics out of traditional natural philosophy and made it in an exact science.

Faraday, Michael (1791-1867), was a British physicist and chemist who is regarded as one of the greatest experimental scientists of the 1800s. He made pioneering contributions to electricity, inventing the electric motor, electric generator and transformer and discovering electromagnetic induction and the laws of electrolysis. He also discovered benzene and was the first to observe that the plane of polarization of light is rotated in a magnetic field (Faraday rotation).

His father was a poor blacksmith who went to London to seek work when Faraday was born in 1791. Faraday received only a rudimentary education as a child and although he was literate, he gained little knowledge of mathematics. At the age of 14, he became an apprentice to a bookbinder in London and began to read voraciously. In 1810 Faraday was introduced to the City Philosophical Society and there received a basic grounding in science. He also attended the Royal Institution, where he was enthralled by the lectures and demonstrations by Humphry Davy (1778-1829). In 1812 Faraday came to the end of his apprenticeship and prepared to devote himself to his trade, not expecting to make a career in science. Almost immediately, however, there came an extraordinary stroke of luck. Davy was temporarily blinded by an explosion in a chemistry experiment and asked Faraday to help him until he gained his sight. When he recovered, Faraday sent Davy the finely bound notes of his lectures. Impressed by he young man, Davy marked out as his next permanent assistant at the Royal Institution and Faraday took up this post in 1813.

When Hans Oersted (1777-1851) discovered that a current of electricity flowing through a wire deflected a compass needle, Faraday was asked to investigate the phenomenon by the editor of the Philosophical magazine. Faraday conceived that circular lines of magnetic force are produced around the wire to explain the orientation of Oersted's compass needle, and therefore set about devising an apparatus that would demonstrate this by causing a magnet to revolve around an electric current. He succeeded in October

1821 with an elaborate device consisting of two vessels of mercury connected to a battery. In this experiment, Faraday demonstrated the basic principles governing the electric motor and although practical motors subsequently developed had a very different form from Faraday's apparatus, he is nevertheless usually credited with the invention of the electric motor.

Faraday's conviction that an electric current gave rise to lines of magnetic force arose from his idea that electricity was a form of vibration and not a moving fluid. He believed that electricity was a state of varying strain in the molecules of the conductor, and this gave rise to a similar strain in the medium surrounding the conductor. It was reasonable to consider therefore that the transmitted strain might set up a similar strain in the molecules of another nearby conductor — that a magnetic field might bring about an electric current in the reverse of the electromagnetic effect discovered by Oersted. Faraday eventually succeeded in producing induction in 1831.

Faraday thus is credited with the simultaneous discovery of electromagnetic induction, though the same discovery had been made in the same way by Joseph Henry (1797-1878) in 1830. However, busy teaching, Henry had not been able to publish his findings before Faraday did so, although both men are now credited with the independent discovery of induction.

Faraday's insight enabled him to make another great discovery soon afterwards. In October 1831 he built the first electric generator. This consisted of a copper disc that was rotated between the poles of a magnet; Faraday touched wires to the edge and center of the disc and connected them to a galvanometer, which registered a steady current. This was the first electric generator, and generators employing coils and magnets in the same way as modern generators were developed by others over the next two years.

Faraday also considered the nature of light and in 1846 arrived at a form of the electromagnetic theory of light that was later developed by James Clerk Maxwell (1831-1879). In a brilliant demonstration of both his intuition and foresight, Faraday said "The view which I am so bold to put forth considers radiation as a high species of vibration in the lines of force which are known to connect particles, and also masses of matter, together. It endeavors to dismiss the ether but not the vibrations." It was a bold view, for no scientist until Albert Einstein (1879-1955) was to take such a daring step.

Michael Faraday was a scientific genius of a most extraordinary kind. Without any mathematical training at all, he succeeded in making the basic discoveries on which virtually all our uses of electricity depend and also in conceiving the fundamental nature of magnetism and, to a degree, of electricity and light. He owed this extraordinary degree of insight to an amazing talent for producing valid pictorial interpretations of the workings of nature. Faraday himself was a modest man, content to serve science as best he could without undue reward, and he declined both a knighthood and the Presidency of the Royal Society. Characteristically, he also refused to take part in the preparation of poison gas for use in the Crimean War. His many achievements are honored in the use of his name in science, the farad being the SI unit of capacitance and the Faraday constant being the quantity of electricity required to liberate a standard amount of substance in electrolysis. (1 Faraday, equivalent to 96,500 coulombs, electroplates one mole of singly ionized metal, for example.)

Gauss, Carl Friedrich (1777-1855), was a German mathematician and physicist. Although known mainly for his pioneering work in mathematics, Gauss made important contributions to physics in the investigation of terrestrial magnetism and the derivation of magnetic units.

Gauss was born into a very poor family in Brunswick on 30 April 1777. He was an infant prodigy, able to calculate even before he could talk. Gauss taught himself to read and must also have worked at arithmetic on his own, for as soon as he was sent to school at the age of eight, he presented a brilliantly simple solution to the problem of adding up any set of whole numbers. Gauss was encouraged to develop his mathematical skills at his first school and then at the Gymnasium, which he entered in 1788. He became so proficient that in 1792, at the age of 14, Gauss was given a stipend by the Duke of Brunswick so that he could devote himself to science. He then spent three years at the Brunswick Collegium Carolinum and then in 1795 went on the University of Gottingen.

In mathematics, Gauss made major contributions to number theory and worked out the basis of non-Euclidean geometry and non-commutative algebra. In 1796 he constructed a seventeen side regular polygon using only a ruler and compass, the first construction of a regular figure in modern times. Gauss was also interested in astronomy, and in 1801 he made a remarkable prediction of the position at which the asteroid Ceres could be found. The asteroid, the first to be discovered, was exactly where Gauss had computed it to be. This was done with Gauss' theory of least squares.

Gauss' work at Gottingen Observatory involved him in important work in geodesy from 1817 onwards. He developed new and improved surveying methods, inventing the heliotrope in 1821 to reflect an image of the Sun to mark positions over long distances. From 1831, Gauss extended this work into the field of terrestrial magnetism in collaboration with Wilhelm Weber (1804-1891), carrying out a worldwide magnetic survey from which Gauss in 1839 was able to derive a mathematical formula expressing magnetism at any location.

Gauss and Weber also investigated electromagnetism and in 1833 constructed a moving needle telegraph similar to that later developed by Charles Wheatstone (1802-1875). Gauss did not succeed in gaining support for the development of his telegraph and it was abandoned. He did make a lasting contribution to physics, however, with his insistence in 1833 that all units should be assembled from a few basic or absolute units, specifically length, mass and time. Gauss and Weber also went on to derive units for magnetism, and the CGS unit for magnetic flux density was called the gauss. It has now been replaced by the tesla in the SI system, which does however derive a full consistent set of units from seven basic units in accordance with Gauss' ideas.

Heaviside, Oliver (1850-1925), was a British physicist and electrical engineer who predicted the existence of the ionosphere, which was known for a time as the Kennely-Heaviside Layer. Heaviside made significant discoveries concerning the passage of electrical waves through the atmosphere and along cables, and added the concepts of inductance, capacitance and impedance to electrical science.

Heaviside was born in Camden town, London, on 18 May 1850. His uncle was Charles Wheatstone (1802-1975), who was a pioneer of the telegraph and may have stimulated in him an interest in electricity. Heaviside received only an elementary education, and was mainly self-taught. He took up employment with the Great Northern Telegraph Company at Newcastle-upon-Tyne in 1870, when he was 20. Heaviside never obtained an academic position but received several honors, including fellowship of the Royal Society in 1891 and the first award of the Faraday Medal by the Institution of Electrical Engineers shortly before he died in Paignton, Devon, on 3 February 1925.

During his early years, Heaviside was absorbed by mathematics. He was familiar with the attempt by Peter Tait (1831-1901) to popularize the quaternions of William Rowan Hamilton (1805-1865), but he rejected the scalar part of the quaternions in his notion of a vector. Heaviside's vectors were of the form $\mathbf{v} = a\mathbf{i} + b\mathbf{j} + c\mathbf{k}$, where $\mathbf{i}$, $\mathbf{j}$ and $\mathbf{k}$ are unit vectors along the Cartesian x-, y- and z-axes respectively. When Heaviside became involved with work relating to Maxwell's famous theory of electricity, he wrote Maxwell's equations in vector form incorporating some discoveries of his own. Heaviside also used operator techniques in his expression of calculus and made much use of divergent series (those with terms whose sum does not approach a fixed amount). He made great use of these in his electrical calculations when many mathematicians were afraid to venture away from convergent series.

When Heaviside became involved with the passage of electricity along conductors, he modified Ohm's law to include inductance and this, together with other electrical properties, resulted in his derivation of the equation of telegraphy. On considering the problem of signal distortion in a telegraph cable, he came to the conclusion that this could be substantially reduced by the addition of small inductance coils throughout its length, and this method has since been used to great effect.

Heaviside considered wireless telegraphy and, in drawing attention to the enormous power required to send useful signals long distances, he suggested that they may be guided by hugging the land and sea. He also suggested that part of the atmosphere may act as a good reflector of electrical waves. Arthur Kennelly (1861-1939) in the United States also made a similar suggestion and this layer, which was for some time known as the Kennelly-Heaviside layer, was subsequently shown to exist by Edward Appleton (1892-1965). This part of the atmosphere, about 100 km above the ground, is now known as the ionosphere.

Helmholtz, Hermann Ludwig Ferdinand von (1821-1894), was a German physicist and physiologist who made a major contribution to physics with the first precise formulation of the principle of conservation of energy. He also made important advances in physiology, in which he first measured the speed of nerve impulses, invented the opthalmoscope and revealed the mechanism by which the ear senses tone and pitch. Helmholtz also made important discoveries in physical chemistry.

APPENDIX B

Helmholtz was born in Potsdam on 31 August 1821. His father was a teacher of philosophy and literature at the Potsdam Gymnasium and his mother was a descendant of William Penn, the founder of the state of Pennsylvania in the United States. Although a delicate child, Helmholtz thrived on scholarship of which his home was not short; his father taught him Latin, Greek, Hebrew, French, Italian and Arabic. He attended the local Gymnasium and showed a talent for physics, but as his father could not afford a university education for him, Helmholtz embarked on the study of medicine, for which financial aid was available in return for a commitment to serve as an army doctor for eight years. In 1838 Helmholtz entered the Friedrich Wilhelm Institute in Berlin, where his time was spent not only on medical and physiological studies but also on chemistry and higher mathematics. He also became an expert pianist.

In 1842 Helmholtz received his MD and became a surgeon with his regiment at Potsdam, where he carried out experiments in a laboratory he set up in the barracks. His ability in science as opposed to medicine was soon recognized, and in 1848 he was released from his military duties. The following year, he took up the post of Associate Professor of Physiology at Konigsberg. In 1855 Helmholtz moved to Bonn to become Professor of Anatomy and Physiology, and then in 1858 was appointed Professor of Physiology at the University of Heidelberg. In 1871 he took up the Chair of Physics at the University of Berlin, and in 1887 became Director of the new Physico-Technical Institute of Berlin. His health then began to fail, and Helmholtz died at Berlin on 8 September 1894.

Helmholtz's doctoral submission in 1842 was concerned with the relationship between the nerve fibers and nerve cells of invertebrates, which led to an investigation of the nature and origin of animal heat. In 1850 he measured the velocity of nerve impulses by experiments with a frog and found it to be about a tenth of the speed of sound. He went on to investigate muscle action and found in 1848 that animal heat and muscle action are generated by chemical changes in the muscles.

This work led Helmholtz to the idea of the principle of conservation of energy, observing that the energy of life processes is derived entirely from oxidation of food. His deduction was remarkably similar to that made by Julius Mayer (1814-1878) a few years earlier, but Helmholtz was not aware of Mayer's work and when he did arrive at a formulation of the principle in 1847, he expressed it in a more effective way. Helmholtz also had the benefit of the precise experimental determination of the mechanical equivalent of heat published by James Joule (1818-1889) in 1847. He was able to derive a general equation that expressed the kinetic energy of a moving body as being equal to the product of the force and distance through which the force moves to bring about the energy change. This equation could be applied in many fields to show that energy is always conserved and it led to the first law of thermodynamics, which states that the total energy of a system and its surroundings remains constant even if it may be changed from one form of energy to another.

Helmholtz went on to develop thermodynamics in physical chemistry, and in 1882 he derived an expression that relates the total energy of a system to its free energy (which is the proportion that can be converted to forms other than heat) and to its temperature and entropy. It enables chemists to determine the direction of a chemical reaction. In this work, Helmholtz was anticipated independently by Willard Gibbs (1839-1903) and both scientists are usually given credit for it.

Helmholtz's work on nerve impulses led him also to make important discoveries in the physiology of vision and hearing. He invented the ophthalmoscope, which is used to examine the retina, in 1851 and the ophthalmometer, which measures the curvature of the eye, in 1855. He also revived the three-color theory of vision first proposed in 1801 by Thomas Young (1773-1829), who like Helmholtz was also a physician and physicist.

Helmholtz dominated German science during the mid 1800s, his wide-ranging and exact work bringing it to the forefront of world attention, a position Germany was to enjoy well into the following century. He served as a great inspiration to others, not least his many students. Foremost of these was Heinrich Hertz (1857-1894), who as a direct result of Helmholtz's encouragement discovered radio waves. Helmholtz took classical mechanics to its limits in physics, paving the way for the radical departure from tradition that was soon to follow with the quantum theory and relativity. When this revolution did occur, however, it was ushered in mainly by German scientists applying the mathematical and experimental expertise upon which Helmholtz had insisted.

Henry, Joseph (1797-1878), was an American physicist who carried out early experiments in electromagnetic induction.

Henry was born in Albany, New York, on 17 December 1797, the son of a laborer. He had little schooling, working his way through Albany Academy to study medicine and then engineering (from 1825). He worked at the academy as a teacher, and in 1826 was made Professor of Mathematics and Physics. He moved to New Jersey College (later Princeton) in 1832 as Professor of Natural Philosophy, in which post he lectured in most of the sciences. He was the Smithsonian Institution's first Director (1846) and first President of the National Academy of Sciences (1868), a position he held until his death, in Washington, on 13 May 1878.

Many of Henry's early experiments were with electromagnetism. By 1830 he had made powerful electromagnets by using many turns of fine insulated wire wound on iron cores. In that year he anticipated Michael Faraday's discovery of electromagnetic induction (although Faraday published first), and two years later he discovered self-induction. He also built a practical electric motor and in 1835 developed the relay (later to be much used in electric telegraphy). In astronomy Henry studied sunspots and solar radiation, and his meteorological studies at the Smithsonian led to the founding of the US Weather Bureau. The unit of inductance was named the henry in 1893.

Hertz, Heinrich Rudolf (1857-1894), was a German physicist who discovered radio waves. His name is commemorated in the use of the hertz as a unit of frequency, one hertz being equal to one complete vibration or cycle per second.

Hertz was born in Hamburg on 22 February 1857. As a child he was interested in practical things and equipped his own workshop. At the age of 15 he entered the Johanneum Gymnasium and, on leaving school three years later, went to Frankfurt to gain practical experience as the beginning of a career in engineering. Engineering qualifications were governed by a state examination, and he went to Dresden Polytechnic to work for this in 1876. During a year of compulsory military service from 1876 to 1877, Hertz decided to be a scientist rather than an engineer and on his return he entered Munich University. He began studies in mathematics, but soon switched to practical physics, which greatly interested him.

Hertz moved to Berlin in 1878 and there came into contact with Herman Helmholtz (1821-1894), who immediately recognized his talents and encouraged him greatly. He gained his Ph.D. in 1880 and remained at Berlin for a further three years to work with Helmholtz as his assistant. He carried out his most important work, discovering radio waves in 1888. In the same year, he began to suffer from toothache, the prelude to a long period of increasingly poor health due to a bone disease. Hertz moved to Bonn in 1889 as Professor of Physics. His physical condition steadily deteriorated and he died of blood poisoning on 1 January 1894, at the early age of 36.

In 1879 a prize was offered by the Berlin Academy for the solution of a problem concerned with Maxwell's theory of electricity, which was set by Helmholtz with Hertz particularly in mind. In 1885 that Hertz found the facilities and the time to work on the problem set by the Berlin Academy. He used large improved induction coils to show that dielectric polarization leads to the same electromagnetic effects as do conduction currents; and he generally confirmed the validity of the famous equations of James Clerk Maxwell concerning the behavior of electricity. In 1888 he realized that Maxwell's equations implied that electric waves could be produced and would travel through air. Hertz went on to confirm his prediction by constructing an open circuit powered by an induction coil, and an open loop of wire as a receiving circuit. As a spark was produced by the induction coil, so one occurred in the open receiving loop. Electric waves traveled from the coil to the loop, creating a current in the loop and causing a spark to form across the gap. Hertz went on to determine the velocity of these waves (which were later called radio waves), and on showing that it was the same as that of light.

From about 1890 Hertz gained an interest in mechanics. He developed a system with only one law of motion — that the path of a mechanical system through space is as straight as possible and is traveled with uniform motion. When this law is developed subject to the nature of space and the constraints on matter, it can be shown that the mechanics of Isaac Newton (1642-1727), Joseph Lagrange (1736-1812) and William Rowan Hamilton (1805-1865), who extended the methods of dealing with mechanical systems considerably, arise as special cases.

Heinrich Hertz made a discovery vital to the progress of technology by demonstrating the generation of radio waves. It is tragic that his early death robbed him of the opportunity to develop his achievements and to see Guglielmo Marconi (1874-1937) and others transform his discovery into a worldwide method of communication.

Kirchhoff, Gustav Robert (1824-1887), was a German physicist who founded the science of spectroscopy. He also discovered laws that govern the flow of electricity in electrical networks and the absorption and emission of radiation in material bodies.

Kirchhoff was born at Konigsberg, Germany (now Kaliningrad, USSR) on 12 March 1824. He studied at the University of Konigsberg, graduating in 1847. In the following year, he became a lecturer at Berlin and in 1850 was appointed Extraordinary Professor of Physics at Breslau. Robert Bunsen (1811-1899) went to Breslau the following year and began a fruitful collaboration with Kirchhoff. In 1852 Bunsen moved to Heidelberg and Kirchhoff followed him in 1854, becoming Ordinary Professor of Physics there. Kirchhoff stayed at Heidelberg until 1875, when he moved to Berlin as Professor of Mathematical Physics. Illness forced him to retire in 1886 and he died at Berlin on 17 October 1887.

Kirchhoff made his first important contribution to physics while still a student. In 1845 and 1846 he extended Ohm's law to networks of conductors and derived the laws known as Kirchhoff's laws that determine the value of the current and potential at any point in a network. He went on to consider electrostatic charge and in 1849 showed that electrostatic potential is identical to tension, thus unifying static and current electricity. Kirchhoff made another fundamental discovery in electricity in 1857 by showing theoretically that an oscillating current is propagated in a conductor of zero resistance at the velocity of light. This was important in the development in the 1860s of the electromagnetic theory of light by James Clerk Maxwell (1831-1879) and Ludwig Lorenz (1829-1891).

Kirchhoff also made another important discovery in 1859 while investigating spectroscopy as an analytical tool. He noticed that certain dark lines in the Sun's spectrum, which had been discovered by Joseph Fraunhofer (1787-1826), were intensified if the sunlight passed through a sodium flame. This observation had in fact been made by Jean Foucault (1819-1868) ten years earlier, but he had not followed it up. Kirchhoff immediately came to the correct conclusion that the sodium flame was absorbing light from the sunlight of the same color that it emitted, and explained that the Fraunhofer lines are due to the absorption of light by sodium and other elements present in the Sun's atmosphere.

Kirchhoff went on to identify other elements in the Sun's spectrum in this way, and also developed the theoretical aspects of this work. In 1859 he announced another important law which states that the ratio of the emission and absorption powers of all material bodies is the same at a given temperature and a given wavelength of radiation produced. From this, Kirchhoff went on in 1862 to derive the concept of a perfect black body — one that would absorb and emit radiation at all wavelengths.

Kirchhoff's contributions to physics had far-reaching practical and theoretical consequences. The discovery of spectroscopy led to several new elements and to methods of determining the composition of stars and the structure of the atom. The study of blackbody radiation led directly to the quantum theory and a radical new view of the nature of matter.

Lorentz, Hendrik Antoon (1853-1928), was a Dutch physicist who helped to develop the theory of electromagnetism, which was recognized by the award (jointly with his pupil Pieter Zeeman) of the 1902 Nobel Prize in Physics.

Lorentz was born in Arnherm, Holland, on 18 July 1853. He was educated at local schools and at the University of Leyden, which he left at the age of 19 to return to Arnherm as a teacher while writing his Ph.D. thesis on the theory of light reflection and refraction. By the time he was 24 he was Professor of Theoretical Physics at Leyden. He remained there for 39 years, before taking up the Directorship of the Teyler Institute in Haarlem where he was able to use the museum's laboratory facilities. He died in Haarlem on 4 February 1928.

Much of Lorentz's work was concerned with James Clerk Maxwell's theory of electromagnetism and his development of it became fundamental to Albert Einstein's Special Theory of Relativity. Lorentz attributed the generation of light by atoms to oscillations of charged particles (electrons) within them. This theory was confirmed in 1896 by the discovery of the Zeeman effect, in which a magnetic field splits spectral lines.

In 1904 Lorentz extended the work of George Fitzgerald to account for the negative result of Michelson-Morley experiment and produced the so-called Lorentz transformations, which mathematically predict the changes to mass, length and time for an object travelling at near the speed of light.

Maxwell, James Clerk (1831-1879), was a British physicist who discovered that light consists of electromagnetic waves and established the kinetic theory of gases. He also proved the nature of Saturn's rings and demonstrated the principles governing color vision.

Maxwell was born at Edinburgh on 13 November 1831. He was educated at Edinburgh Academy from 1841 to 1847, when he entered the University of Edinburgh. He then went on to study at Cambridge in 1850, graduating in 1854. He became Professor of Natural Philosophy at Marischal College, Aberdeen, in 1856 and moved to London in 1860 to take up the post of Professor of Natural Philosophy and Astronomy at King's College. On the death of his father in 1865, Maxwell returned to research. However in 1871 he was persuaded to move to Cambridge, where he became the first Professor of Experimental Physics and set up the Cavendish Laboratory, which opened in 1874. Maxwell continued in this position until 1879, when he contracted cancer. He died at Cambridge on 5 November 1879, at the early age of 48.

Maxwell demonstrated his great analytical ability at the age of 15, when he discovered an original method for drawing a perfect oval. His first important contribution to science was made from 1849 onwards, when Maxwell applied himself to color vision. He revived the three-colored theory of Thomas Young (1773-1829) and extended the work of Hermann Helmholtz (1821-1894) on color vision. Maxwell showed how colors could be built up from mixtures of the primary colors red, green and blue, by spinning discs containing sectors of these colors in various sizes. In the 1850s, he refined this approach by inventing a color box in which the three primary colors could be selected from the Sun's spectrum and combined together. Maxwell confirmed Young's theory that the eye has three kinds of receptors sensitive to the primary colors and showed that color blindness is due to defects in the receptors. He also explained fully how the addition and subtraction of primary colors produces all other colors, and crowned this achievement in 1861 by producing the first color photograph to use a three-color process. This picture, the ancestor of all color photography, printing and television, was taken of a tartan ribbon by using red, green and blue filters to photograph the tartan and to project a colored image.

Maxwell worked on several areas of inquiry at the same time, and from 1855 to 1859 took up the problem of Saturn's rings. No one could give a satisfactory explanation for the rings that would result in a stable structure. Maxwell proved that a solid ring would collapse and a fluid ring would break up, but found that a ring composed of concentric circles of satellites could achieve stability, arriving at the correct conclusion that the rings are composed of many small bodies in orbit around Saturn.

Maxwell's development of the electromagnetic theory of light took many years. It began with the paper On Faraday's Lines of Force (1855-1856), in which Maxwell built on the views of Michael Faraday (1791-

1867) that electric and magnetic effects result from fields of lines of force that surround conductors and magnets. Maxwell drew an analogy between the behavior of the lines of force and the flow of an incompressible liquid, thereby deriving equations that represented known electric and magnetic effects. The next step towards the electromagnetic theory took place with the publication of the paper On Physical Lines of Force (1861-1862). Here Maxwell developed a model for the medium in which electric and magnetic effects could occur. This could throw some light on the nature of lines of force. He devised a hypothetical medium consisting of an incompressible fluid containing rotating vortices responding to magnetic intensity separated by cells responding to electric current.

By considering how the motion of the vortices and cells could produce magnetic and electric effects, Maxwell was successful in explaining all known effects of electromagnetism, showing that the lines of force must behave in a similar way. However Maxwell went further, and considered what effects would be caused if the medium were elastic. It turned out that the movement of a charge would set up a disturbance in the medium, forming transverse waves that would be propagated through the medium. The velocity of these waves would be equal to the ratio of the value for a current when measured in electrostatic units and electromagnetic units. This had been determined by Friedrich Kohlrausch (1840-1910) and Wilhelm Weber (1804-1891), and it was equal to the velocity of light. Maxwell thus inferred that light consists of transverse waves in the same medium that causes electric and magnetic phenomena.

Maxwell was reinforced in this opinion by work undertaken to make basic definitions of electric and magnetic quantities in terms of mass, length and time. In On the Elementary Regulations of Electric quantities (1863), he found that the ratio of the two definitions of any quantity based on electric and magnetic forces is always equal to the velocity of light. He considered that light must consist of electromagnetic waves, but first needed to prove this by abandoning the vortex analogy and arriving at an explanation based purely on dynamic principles. This he achieved in A Dynamical Theory of the Electromagnetic Field (1864), in which he developed the fundamental equations that describe the electromagnetic field. These showed that light is propagated in two waves, one magnetic and the other electric, which vibrate perpendicular to each other and to the direction of propagation. This was confirmed in Maxwell's Note On The Electromagnetic Theory of Light (1868), which used an electrical derivation of the theory instead of the dynamical formulation, and Maxwell's whole work on the subject was summed up in Treatise On Electricity and Magnetism in 1873.

The treatise also established that light has a radiation pressure, and suggested that a whole family of electromagnetic radiations must exist of which light was only one. This was confirmed in 1888 with the sensational discovery of radio waves by Heinrich Hertz (1857-1894). Sadly, Maxwell did not live long enough to see this triumphant vindication of his work. He also did not live to see the ether (the medium in which light waves were said to be propagated) disproved with the classic experiments of Albert Michelson (1852-1931) and Edward Morley (1838-1923) in 1881 and 1887, which Maxwell himself suggested in the last year of his life. However this did not discredit Maxwell as his equations and description of electromagnetic waves remain valid even though the waves require no medium

Maxwell's other major contribution to physics was to provide a mathematical basis for the kinetic theory of gases. Here he built on the achievements of Rudolf Clausius (1822-1888), who in 1857 to 1858 had shown that a gas must consist of molecules in constant motion colliding with each other and the walls of the container. Clausius developed the idea of the mean free path, which is the average distance that a molecule travels between collisions. As the molecules have a high velocity, the mean free path must be very small, otherwise gases would diffuse much faster than they do and have greater thermal conductivities.

Maxwell's development of the kinetic theory was stimulated by his success in the similar problem of Saturn's rings. It dates from 1860, when he used a statistical treatment to express the wide range of velocities that the molecules in a quantity of gas must inevitably possess. He arrived at a formula to express the distribution of velocity in gas molecules, relating it to temperature and thus finally showing that heat resides in the motion of molecules — a view that had been suspected for some time. Maxwell then applied it with some success to viscosity, diffusion and other properties of gases that depend on the nature of the motion of their molecules.

However, in 1865, Maxwell and his wife carried out exacting experiments to measure the viscosity of gases over a wide range of pressure and temperature. They found that the viscosity is independent of the pressure and that it is very nearly proportional to the absolute temperature. This later finding conflicted with the previous distribution law and Maxwell modified his conception of the kinetic theory by assuming that molecules do not undergo elastic collisions as had been thought but are subject to a repulsive force that varies inversely with the fifth power of the distance between them. This led to new equations that satisfied the viscosity-temperature relationship as well as the laws of partial pressures and diffusion.

However, Maxwell's kinetic theory did not fully explain heat conduction, and it was modified by Ludwig Boltzmann (1844-1906) in 1868, resulting in the Maxwell-Boltzmann distribution law. Both men thereafter contributed to successive refinements of the kinetic theory and it proved fully applicable to all properties of gases. It also led Maxwell to an accurate estimate of the size of molecules and to a method of separating gases in a centrifuge. The kinetic theory, being a statistical derivation, also revised opinions on the validity of the second law of thermodynamics, which states that heat cannot of its own accord flow from a colder to a hotter body. In the case of two connected containers of gases at the same temperature, it is statistically possible for the molecules to diffuse so that the faster-moving molecules all concentrate in one container while the slower molecules gather in the other, making the first container hotter and the second colder. Maxwell conceived this hypothesis, which is known as Maxwell's demon. Even though this is very unlikely, it is not impossible and the second law can therefore be considered to be not absolute but only highly probable.

Maxwell is generally considered to be the greatest theoretical physicist of the 1800s, as his forebear Faraday was the greatest experimental physicist. His rigorous mathematical ability was combined with great insight to enable him to achieve brilliant syntheses of knowledge in the two most important areas of physics at that time. In building on Faraday's work to discover the electromagnetic nature of light, Maxwell not only explained electromagnetism but also paved the way for the discovery and application of the whole spectrum of electromagnetic radiation that has characterized modern physics. In developing the kinetic theory of gases, Maxwell gave the final proof that the nature of heat resides in the motion of molecules.

Oersted, Hans Christian (1777-1851), was a Danish physicist who discovered that an electric current produces a magnetic field.

Oersted was born at Rudkobing, Langeland, on 14 august 1777. He had little formal education as a child but on moving to Copenhagen in 1794, Oersted entered the university and gained a degree in pharmacy in 1797, proceeding to a doctorate in 1799. He then worked as a pharmacist before making a tour of Europe from 1801 to 1803 to complete his studies in science.

Oersted made his historic discovery of electromagnetism in 1820, but he had been seeking the effect since 1813 when he predicted that an electric current would produce magnetism when it flowed through a wire just as it produced heat and light. Oersted made this prediction on philosophical grounds, believing that all forces must be interconvertible. Others were also seeking the electromagnetic effect, but considered that the magnetic field would lie in the direction of the current and had not been able to detect it. Oersted reasoned that the effect must be a lateral one and early in 1820, he set up an experiment with a compass needle placed beneath a wire connected to a battery. A lecture intervened before he could perform the experiment and, unable to wait, Oersted decided to try it out before his students. The needle moved feebly, making no great impression on the audience but thrilling Oersted. Because the effect was so small, Oersted delayed publication and investigated it more fully, finding that a circular magnetic field is produced around a wire carrying a current. He communicated this momentous discovery to the major scientific journals of Europe in July 1820.

In 1822, Oersted turned to the compressibility of gases and liquids, devising a useful apparatus to determine compressibility. He also investigated thermoelectricity, in 1823. Oersted may have wanted to continue work on electromagnetism, but his sensational discovery resulted in an explosion of activity by other scientists and Oersted possibly felt unable to compete. Major theoretical and practical advances were made by Andre Ampere (1775-1836) and Michael Faraday (1791-1867) soon afterwards, Oersted thereby providing the basis for the main thrust of physics in the 1800s.

Volta, Alessandro (1745-1827), was an Italian physicist who discovered how to produce electric current and built the first electric battery. He also invented the electrophorous as a ready means of producing charges of static electricity.

Volta was born at Como on 18 February 1745. He received his early education at various religious institutions but showed a flair for science, particularly the study of electricity which had been brought to the forefront of attention by the experiments and theories of Benjamin Franklin (1706-1790). Volta began to experiment with static electricity in 1765 and soon gained a reputation as a scientist, leading to his appointment as principal of the Gymnasium at Como in 1774 and, a year later, as Professor of Experimental Physics there.

Volta's first major contribution to physics was the invention of the electrophorus in 1775. He had researched thoroughly into the nature and quantity of electrostatic charge generated by various materials, and he used this knowledge to develop a simple practical device for the production of charges. His electrophorus consisted of a disc made of turpentine, resin and wax, which was rubbed to give it a negative charge. A plate covered in tin foil was lowered by an insulated handle on to the disc, which induced a positive charge that was likewise induced on the upper surface was removed by touching it to ground the charge, leaving a positive charge on the foil. This process could then be repeated to build up a greater and greater charge. Volta went on to realize from his electrostatic experiments that the quantity of charge produced is proportional to the product of its tension and the capacity of the conductor. He developed a simple electrometer similar to the gold-leaf electroscope but using straws so that it was much cheaper to make. This instrument was very sensitive and Volta was able to use it to measure tension, proposing a unit that was equivalent to about 13,500 volts.

Volta's greatest contribution to science began with the discovery by Luigi Galvani (1737-1798) in 1791 that the muscles in dead frogs contract when two dissimilar metals (brass and iron) are brought into contact with the muscle and each other. Volta successfully repeated Galvani's experiments using different metals and different animals, and he also found that placing the two metals on his tongue produced an unpleasant sensation. The effects were due to electricity and in 1792, Volta concluded that the source of the electricity was in the junction of the metals and not, as Galvani thought, in the animals. Volta even succeeded in producing a list of metals in order of their electricity production based on the strength of the sensation they made on his tongue, thereby deriving the electromotive series.

Volta's tongue and Galvani's frogs proved to be highly sensitive detectors of electricity — much more so than Volta's electrometer. In 1796, Volta set out to measure the electricity produced by different metals, but to register any deflection in the electrometer he had to increase the tension by multiplying that given by a single junction. He soon hit upon the idea of piling discs of metals on top of each other and found that they had to be separated by a moist conductor to produce a current. The political upheavals of this period prevented Volta from proceeding immediately to construct a battery but he had undoubtedly achieved the "voltaic pile", as it came to be called, by 1800. In that year, he wrote to the President of the Royal Society, Joseph Banks (1743-1820), and described two arrangements of conductors that produced an electric current. One was a pile of silver and zinc discs separated by cardboard moistened with brine, and the other a series of glasses of salty or alkaline water in which bimetallic curved electrodes were dipped.

Volta's discovery was a sensation, for it enabled high electric currents to be produced for the first time. It was quickly applied to produce electrolysis, resulting in the discovery of several new chemical elements, and then led throughout the 1800s to the great discoveries in electromagnetism and electronics which culminated in the invention of the electrical machines and electronic devices that we enjoy today. Volta's genius lay in an ability to construct simple devices and in his tenacity to follow through his convictions. He was not a great theoretician and did not attempt to explain his discovery. However, he did see the need for establishing proper measurement of electricity, and it is fitting that the unit of electric potential, tension or electromotive force is named the volt in his honor.

Weber, Wilhelm Eduard (1804-1891), was a German physicist who made important advances in the measurement of electricity and magnetism by devising sensitive instruments and defining electric and magnetic units. In recognition of his achievements, the SI unit of magnetic flux density is called the weber. Weber was also the first to reach the conclusion that electricity consists of charged particles.

Weber was born at Wittenberg on 24 October 1804. The family was highly gifted, for Weber's father was Professor of Theology at the University of Wittenberg and his older brother Ernst Weber (1795-1878) became a pioneer in the physiology of perception. In 1814, the family moved to Halle, where Weber entered the University in 1822. He obtained his doctorate in 1826, and then became a lecturer at Halle, rising to an assistant professorship in 1828.

In 1831, Weber moved to Gottingen to become Professor of Physics and there began a close collaboration with Carl Gauss (1777-1855). He lost this post in 1837 following a protest at the suspension of the constitution by the new ruler of Hanover. However Weber managed to continue working at Gottingen and then in 1843 he obtained the position of Professor of Physics at Leipzig. In 1849, Weber returned to his former post at Gottingen. He remained in this position until he retired in the 1870s, and died at Gottingen on 23 June 1891.

Weber's work in magnetism dates from the time when he joined Gauss at Gottingen. They conceived absolute units of magnetism that were defined by expressions involving only length, mass and time, and Weber went on to construct highly sensitive magnetometers. He also built a 3-km telegraph to connect the physics laboratory with the astronomical observatory where Gauss worked, and this was the first practical telegraph to operate anywhere in the world. It was not subsequently developed as a commercial invention, and from 1836 to 1841 Gauss and Weber organized a network of observation stations to correlate measurements of terrestrial magnetism made around the world.

In 1840, Weber extended his work on magnetism into the realm of electricity. He defined an electromagnetic unit for electric current that was applied to measurements of current made by the deflection of the magnetic needle of a galvanometer. In 1846, he developed the electrodynamometer, in which a current causes a coil suspended within another coil to turn when a current is passed through both. This instrument could be used to measure alternating currents. Current could also be measured by the Coulomb torsion balance in electrostatic units, and in an experiment carried out with Rudolph Kohlrausch in 1855, Weber found that the ratio of the electromagnetic unit to the electrostatic unit is a constant equal to the velocity of light. They did this by discharging a condenser through a torsion balance and a ballistic galvanometer. The precise work by Weber and Kohlrausch was praised by James Clerk Maxwell (1831-1879) as one of the most important steps in the progress of the science. Maxwell later developed the electromagnetic theory of light.

In 1852, Weber defined the absolute unit of electrical resistance and also began to conceive of electricity in terms of moving charged particles of positive and negative electricity, resistance being produced by a combining of the two particles that prevents current flow. In 1846, he had produced a general law of electricity to express electrical effects mathematically as was introduced in Chapter 7. In a remarkable piece of foresight, he put forward in 1871 the view that atoms contain positive charges that are surrounded by rotating negative particles and that the application of an electric potential to a conductor causes the negative particles to migrate from one atom to an other. It was not until 1913, with the proposal of the Rutherford-Bohr model of the atom, that Weber's ideas were seen to be essentially correct. Weber also provided similar explanations of thermal conduction and thermoelectricity that were later fully developed by others.

Weber also did important work in acoustics with his brother Ernst Weber. In 1825, they made the first experimental study of interference in sound. Their findings were important in enabling Hermann Helmholtz (1821-1894) to achieve explanations of the perception of sound and mechanism of hearing.

Weber's insistence on precise experimental work to produce correct definitions of electrical and magnetic units was very important to the development of these sciences and to electromagnetism. His far-reaching views on the nature of electricity were influential in creating a climate for the acceptance of such ideas when evidence for them was later found.

REFERENCES

[1] Weber's theory is best described in; A. K. T. Assis, *Weber's Electrodynamics*, Kluwer Academic Publishers, Dordrecht, The Netherlands (1994).

[2] H. Goldstein, *Classical Mechanics*, Addison-Wesley, Reading, Massachusetts, p. 19 (1950).

[3] H. V. Helmholtz, Philosophical Magazine **44**, 530-537 (1872).

[4] Junichiro Fukai and Earl T. Kinzer, "Compatibility of Weber's force with Maxwell's equations," Galilean Electrodynamics, Vol. **8** (1997), 53-55.

[5] Richard P. Feynman, Robert B. Leighton, and Matthew Sands, *The Feynman Lectures on Physics*, Vol. II, Addison-Wesley, Reading, Massachusetts, p. 26-5 (1964).

[6] D. Corson and P. Lorrain, *Introduction to electromagnetic Fields and Waves*, W. H. Freeman, San Francisco (1962).

[7] J. R. Reitz, F. J. Milford, and R. W. Christy, *Foundations of Electromagnetic Theory*, Third Edition, Addison-Wesley, Reading, Massachusetts (1979).

[8] Hans C. Ohanian, *Classical Electrodynamics*, Allyn and Bacon, Inc, Newton, Massachusetts, p. 256 (1988).

[9] John David Jackson, *Classical Electrodynamics*, Third Edition, John Wiley & Sons, New York, NY, p. 176 (1999).

[10] David J. Griffiths, *Introduction to Electrodynamics*, Third Edition, Prentice-Hall, Upper Saddle River, New Jersey, p. 219 (1999).

[11] Edward M. Purcell, *Electricity and Magnetism*, Second Edition, McGraw-Hill, New York, NY (1985), p. 189.

REFERENCES

[12] By looking at Lorentz contraction, Feynman (Reference 5, Vol. II, 13-6) explains the magnetic force on a positive charge moving with velocity **v** outside a current-carrying wire where the drift velocity is the same as **v**. With respect to the moving charge, the density of the moving electrons is λ, whereas the density of the positive grid of the wire is $\gamma\lambda$, where γ is the Lorentz factor. The result is a net repulsion of the positive charge. Now if we extend the same logic to a stationary charge outside the wire, the same Lorentz-contraction argument applies, but the other way. The electrons, in this case, have the higher density, and there should be a force on our stationary positive charge toward the wire. This was noted by Howard Hayden (private communication).

[13] A. K. T. Assis, W. A. Rodorigues Jr, and A. J. Mania, "The electric field outside a stationary wire carrying a constant current,' Foundation of Physics **29**, 729-853 (1999).

[14] Oleg D. Jefimenko, *Electricity and Magnetism*, Appleton-Century-Crofts, New York, p. 299-419 and 509-611 (1966).

[15] W. F. Edwards, C. S. Kenyon and D. K. Lemon, "Continuing investigation into possible electric fields arising from steady conduction currents," Physical Review D **14**, 922-938 (1976).

[16] Howard Hayden, "Possible explanation of for the Edward effect," Galilean Electrodynamics, Vol. **1**, 33-35 (1990). More references will be found in Ref. 1, p. 168.

[17] P. Graneau, J. Applied physics **62**, 3006-3009. Many references therein.

[18] Yuan Zhong Zhang, *Special relativity and Its Experimental Foundations*, Chapter 11, p. 225, (World Scientific, Singapore, 1997). See also http://www2.slac.stanford.edu/vvc/theory/relativity.html

[19] J. C. Maxwell, *A treatise on Electricity and Magnetism*, Vol. 2, Dover, New York, Arts. [686-688], p. 318-320 (1954).

[20] See a review paper on this subject by P. Graneau," The Ampere-Neumann electrodynamics of metallic conductors." Fortschritte der Physik **34**, 457-503 (1986).

[21] P. Graneau, "Electromagnetic jet-propulsion in the direction of current flow," Nature **295**, 92-96 (1982).

[22] R.A. Clemente and A. K. T. Assis, "Two-body problems for Weber-like interactions", International Journal of Theoretical Physics **30**, 537-545 (1991).

[23] A. E. Chubykalo and R. Smirnov-Rueda, Mod. Phys. Letters **A12**, 1-24 (1997).

[24] W. Weber, Annalen der Physik **73**, 193-240 (1848); J. C. Maxwell, *A treatise on Electricity and Magnetism*, Vol. 2, Dover, New York, Arts. [846-851], p. 480-483 (1954)

[25] E. T . Whittaker, *A History of the Theories of Aether and Electricity*, Humanities Press, New York, Vol. 1, p. 201-203 (1973).

[26] L. Rosenfeld, "The velocity of light and the evolution of electrodynamics," Il Nuovo Cimento, Supplement **4**, 1630-1669 (1957).

[27] G. Kirchhoff, "On the motion of electricity in wires," Philosophical Magazine **13**, 393-412 (1857).

[28] O. Heaviside, Philosophical Magazine **27**, 324 (1889).

[29] W. D. Niven (edited), *The Scientific Papers of James Clerk Maxwell*, Cambridge University Press, Cambridge, 1820, reprinted by Dover, New York (1965), p. 526-527.

[30] E. T. Kinzer and J. Fukai, "Weber's force and Maxwell's equations," Foundations of Physics Letters **67**(8), 574-578 (1997).

[31] F. T. Trouton and H. R. Noble, Phil. Trans. Roy. Soc. London **A202**, 165-181 (1904).

REFERENCES

[32] H. C. Hayden, "High sensitivity Trouton-Noble Experiment," *Rev. Sci. Instr.* **65 (4)** (April 1994).

[33] H. C. Hayden, "Analysis of Trouton-Noble Experiment," Galilean Electrodynamics, Vol. **5**, 83-85 (1994).

[34] Isao Imai, *Electromagnetism Examined* (Japanese), Saiensu-sha, Tokyo (1990), pp. 404-407; A variation is found in D. J. Griffiths, *Introduction to Electrodynamics*, Third edition, Prentice hall, Upper Saddle River (1999) p. 359; For further references, see T. C. E. Ma, Am. J. Phys. **54**, 949 (1986).

[35] Richard P. Feynman, Robert B. Leighton, and Matthew Sands, *The Feynman Lectures on Physics*, Vol. II, Addison-Wesley, Reading, Massachusetts, p. 17-4 (1964).

[36] E. T. Kinzer and J. Fukai, "Induced EMF by Weber's force," Galilean Electrodynamics, Vol. **7**, 39-50 (1996).

[37] A. K. T. Assis, *Weber's Electrodynamics*, Kluwer Academic Publishers, Dordrecht, Germany, P. 187 (1994).

[38] A. K. T. Assis, Journal of the Physical Society of Japan **62**, 1418-1422 (1993).

[39] H. V. Helmholtz, Philosophical Magazine **44**, 530-537 (1872), and Ref. 1, p. 197.

[40] E. Schroedinger, Annalen der Physik **77**, 325-336 (1925).

[41] V. F. Mikhailov, "Action of an electrostatic potential on the electron mass," Annales de la Foundation Louis de Broglie, **24**, 161-169 (1999).

[42] A. K. T. Assis, J. Fukai, and H. B. Carvalho, " Weberian Induction," Physics Letters **A 268**, 274-278 (2000).

[43] A. K. T. Assis, "On the propagation of electromagnetic signals in wires and coaxial cables according to Weber's electrodynamics," Foundations of Physics **30**, 1107-1121 (2000).

[44] Articles about vacuum will be found in: S. Saunders and H. R. Brown edited, *The Philosophy of Vacuum*, Clarendon Press, Oxford (1991); H. C. Von Baeyer, "Vacuum," Discover, March 1992, p. 108-112; F. Wilczek, "The persistence of ether," Physics Today, January 1999, p. 11-13.

[45] Renhe Tian and Zhuhuai Li, "The speed and apparent mass of photons in a gravitational field", Am. J. Phys. **58**, 890-892 (1990).

[46] H. C. Hayden, "Light speed as a function of gravitational potential," Galilean Electrodynamics **1**, 15-17 (1990).

[47] Frank R. Tangherlini, "On Snell's law and the gravitational deflection of light," Am. J. Phys. 36, 1001-1004 (1968).

[48] W. W. Campbell and R. J. Trumpler, "Observations made with a pair of five foot cameras on the light-deflections in the sun's gravitational field at the total solar eclipse of September 21, 1922," Lick Obs. Bull, **13**, 130-160.

[49] C. Moller, *The Theory of Relativity*, Second edition, Clarendon Press, Oxford, London (1972), p. 498.

[50] For a more comprehensive account see Edward G. Harris, *Introduction to Modern Theoretical Physics*, Vol. 1, John Wiley & Sons, New York, NY (1972), Chap. 11.

[51] Every so often a fairly large group of scientists begins to assert that science is just about complete. Just before relativity and quantum mechanics appeared on the scene, the scientists at the turn of the 19th century lamented the fate of their successors for whom only small mop-up operations remained. Said Albert A. Michelson, of Michelson-Morley fame, in a speech at the dedication of Ryerson Physics Labo-

ratory, University of Chicago in 1894, quoted by P. Coveney and R. Highfield in the Arrow of Time, Flamingo, London 19991, p 67:

> "The more important fundamental laws and facts of physical science have all been discovered, and these are now so firmly established that the possibility of their ever being supplanted in consequence of new discoveries is exceedingly remote . . . Our future discoveries must be looked for in the sixth place of decimals."

Among others, in 1900 Lord Kelvin in a lecture said

> "There is nothing new to be discovered in physics now. All that remains is more and more precise measurement."

[52] A. K. T. Assis, *Relational Mechanics*, Apeiron, Montreal, Canada (1999).

[53] E. Mach, *The Science of Mechanics* (Open Court, La Salle, 1960).

[54] V. V. Batygin and I. N. Toptygin, *Problems in Electrodynamics,* Academic Press, London (1964), p. 61.

Reference for Appendix A and B:

For more detailed accounts see David Abbott (Edited), *Physicists*, Peter Bedrick Books, New York, NY (1984).

INDEX

Ampere, Andre-Marie, 11
   Ampere's force law, 37, 40, 55
   Ampere-Maxwell integral equation, 13
   bridge experiment, 41
Aristotle, 79
Assis, Andre K. T., 73, 79, 91
Berkeley, 91
Biot-Savart
   law, 6, 13, 14, 28, 40
Coulomb, Charles, 4, 7
   Coulomb's law, 12, 16, 21, 24, 27, 28, 29, 34, 55, 59, 71.
dark matter, 88
Democritus, 79
Einstein, Albert A., 3, 9
   relativity theory, 3, 88
   theory of gravity, 88
emf
   by Faraday's law, 55
   inferred from Weber force, 62
Faraday, Michael, 1
   Faraday's law, 12, 14, 16, 55, 93
Feynman, Richard P., 68
Foucault's pendulum, 91
Gauss, Carl Friedrich, 59
   Gauss's law, 47
Gilbert, William, 4
Graneau, Peter, 41
   longitudinal force, 41
gravity
   deflection of starlight by, 82
   effect on speed of light, 82
Hayden, Howard C.
   permittivity in varying gravity, 82
   Trouton-Noble experiment, 67
Heaviside, Oliver, 61
Helmholtz, Hermann L. F. von
   argument against Weber's theory, 75
inductance, mutual, 59
induction
   Faraday's law of, 63
   Weberian, 75
Kelvin, Lord, 90
Kirchhoff, Gustav Robert, 60, 79
Lagrangian energy, 65
Leibniz, 91
longitudinal force, 34, 41
Lorentz, Hendrik Autoon, 1
   Lorentz's force, 1, 8, 24, 27, 28, 31, 40, 61, 94
Mach's principle, 91, 92
   and *emf*, 93
MACHOS, 88
Maxwell, James Clerk
   Ampere-Maxwell equation, 12
   compatability of Weber force with Maxwell equations, 62
   longitudinal force, 41
   Maxwell's equations, 11, 56
   Maxwell's equations used to derive Weber force, 59

praise for Ampere's force law, 41
Michell, John, 4
Michelson, A. A., 67, 80, 90
Michelson-Morley experiment, 109
   Trouton-Noble experiment as electrostatic M-M, 67
Mikhailov, V. F., 75
mutual inductance, 59
Newton, Sir Isaac, 3
   third law, 5
Newton's third law, 5, 16, 33
Oersted, Hans Christian, 4
Ohm's law, 60
orbit problem, 49
permittivity
   near massive body, 82, 84
   of free space, 7
refraction, index of, 82
   of gravitaional field, 85
relativity
   general theory of, 88
   general theory of, & curved spacetime, 94
   general theory of, & speed of light, 82
   general theory of, and light bending, 86
   special theory of, 3, 9, 11, 28, 30, 31, 34, 43
   special theory of & Trouton-Noble experiment, 67
   special theory of, contrary to Newton's third law, 33
Rutherford scattering, 49, 51, 72
scattering, 49
SLAC, 35
Snell's law, 87
solenoid, rotating, 68
speed of light, 60, 79, 81
   function of radius, 82
telegraph equation, 60
third law
   Newton's, 5
   Newton's & relativity, 3
Tian and Li
   light in gravitational field, 82
Trouton-Noble experiment, 33, 67
vacuum state, 80
vector potential, 59, 62, 70
Weber, Wilhelm Edward, 59
   electrodynamics applications, 67
   Weber force, 55, 59, 60, 62
   Weber's electrodynamics, 65
   Weber's potential energy, 65
   Weberian induction, 75
WIMPS, 88
zero point energy, 80

About the Author

Junichiro Fukai, an Associate Professor of Physics, has taught physics at Auburn University since 1974.

Born and raised in Japan, he graduated from Waseda University earning a Bachelor of Engineering degree. After working at Toshiba Corporation as an engineer for three years, he came to the United States and entered the University of Denver majoring in Physics and earned his M.S.

He then moved to the University of Tennessee where he earned his Ph.D. in theoretical plasma physics. After postdoctoral studies at the University of Tennessee and Yale University, he went to Auburn University. There he worked on theoretical and experimental studies on plasma instabilities. His interest moved to theoretical investigations of dielectric breakdown phenomena and developing a device to decompose toxic gases by an electric discharge.

At Auburn he taught various physics at almost all levels. Meanwhile, he has been interested in fundamental issues of electrodynamics, relativity, and quantum theory.